Introduction to biodiesel production 1st edition

How to create your own batches and a waste oil processor at home.

Phillip Westinghouse, Delfin

© Copyright 2018 – Alán Adrián Delfín Cota. **All rights reserved.**

The contents of this book may not be reproduced, duplicated or transmitted without direct written permission from the author.

Under no circumstances will any legal responsibility or blame be held against the publisher for any reparation, damages, or monetary loss due to the information herein, either directly or indirectly.

Legal Notice:

This book is copyright protected. This is only for personal use. You cannot amend, distribute, sell, use, quote or paraphrase any part of the content within this book without the consent of the author.

Disclaimer Notice:

Please note the information contained within this document is for educational and entertainment purposes only. Every attempt has been made to provide accurate, up to date and complete, reliable information. No warranties of any kind are expressed or implied. Readers acknowledge that the author is not engaging in the rendering of legal, financial, medical or professional advice. The content of this book has been derived from various sources. Please consult a licensed professional before attempting any techniques outlined in this book.

By reading this document, the reader agrees that under no circumstances is the author responsible for any losses, direct or indirect, which are incurred as a result of the use of information contained within this document, including, but not limited to, —errors, omissions, or inaccuracies.

Dedicatory

To my dad, because he constantly asks me about the content of this book given to his laziness to read.

Contents

CHAPTER 1 ....
- INTRODUCTION ....
- BIODIESEL OVERVIEW ....

BIODIESEL ....
- Why Produce My Own Biodiesel? ....
- Where can I buy biodiesel? ....
- Why is not sold everywhere? ....
- Downsides and benefits of using biodiesel ....
- BIODIESEL BENEFITS ....
- Is it sustainable? ....

CHAPTER 2 ....
- HOW DOES A DIESEL ENGINE WORKS ....
- How Does Bio-Diesel Work? ....

How a four-stroke diesel engine works. ....

CHAPTER 3 ....
- COLLECTING MATERIALS ....

CHAPTER 4 ....
- SETTING UP THE WORKSHOP ....

CHAPTER 5 ....

BASICS TO UNDERSTAND THE CHEMICAL PROCESS ....
- ORGANIC CHEMISTRY BASICS ....

CHAPTER 6 ....
- MAKING BIODIESEL ....
  - The Raw materials For Biodiesel ....

- Water Washing Biodiesel ....
- Biodiesel Safety Tips ....

STEPS TO MAKING BIODIESEL ....

CONCLUSION ....

CHAPTER 1

INTRODUCTION

In today's world, one can hardly escape the subject of fuel prices and fuel supply. For a number of different reasons people have turned from standard petroleum based fuel sources and looked for a reliable alternative-fuel. Biodiesel is one such fuel that experts and enthusiasts have embraced as not only their idea of a fuel of the future, but is also their choice for a fuel for today. But, what is Biodiesel?

Biodiesel is a fuel containing some of the same traits as conventional diesel fuel. Biodiesel is made from high quality vegetable oils through a manufacturing process that can be done on a large scale - such as a refinery, or on a small scale - such as a home Biodiesel kit. The primary use for Biodiesel right now is as a substitute for petroleum based diesel fuel.

In different parts of the world, different plants are used as the source for the vegetable oil that is made into Biodiesel. Theoretically, any vegetable grown can be broken down and turned into Biodiesel, but right now most Biodiesel producers use one primary crop as their source. In America, the primary crop grown for Biodiesel production is corn. Corn, being one of the more common crops grown in American, provides a readily available supply of vegetable matter for Biodiesel manufacturers. In much of Europe, rapeseed is used in the production of Biodiesel. And, in Southeast Asia the primary plant grown for the manufacture of Biodiesel is Soybeans. After going through the manufacturing process, there is little difference in the properties of Biodiesel made from one plant over another.

The current uses for Biodiesel are mainly limited to that of being a cleaner burning replacement for petroleum based diesel fuel. Biodiesel can be both economically viable and highly efficient for most mobile applications. There will be some performance and usage differences between diesel and Biodiesel and they will vary from vehicle to vehicle as expected. It is possible for most modern diesel engines to burn standard Biodiesel. Currently, Biodiesel is also marketed as a Biodiesel/Diesel mix.

BIODIESEL OVERVIEW

Biodiesel is an alternative, the environmental alternative to the fuel extracted by traditional means; biodiesel is the result of a process that uses domestic, renewable supplies. It includes no petroleum, but it can be combined with petroleum diesel, generating a biodiesel mixture. It can be employed in compression-ignition engines with little or no modifications. It is easy to use, ecological, nontoxic, and fundamentally without any amount of aromatics or sulfur.

Biodiesel is obtained through a transesterfication, a chemical process which separates the glycerin from the fat or vegetable oil. The two products this process generates are biodiesel (the scientific chemical equivalent is methyl esters) and glycerin (a derivative substance, usually used in soaps or other beauty products).

How can it be obtained?

The most important aspect is that anyone can make biodiesel, using vegetable oils, animal fats or both!

You can be the "producer" of homemade fuel, which will make you diesel motor work better and last longer. Not to forget that it is much ecological! It will not only be cheaper, but it will help you recycle the garbage.

Transesterfication is fairly straightforward. You don't have to be a chemist to produce biodiesel (despite the fact that it would certainly help!). Many people outside the field of chemistry have sketched consistent and verified biodiesel fabrication recipes. In this procedure the vegetable oil or animal fat comes in reaction with alcohol and a catalyst.

The percentage is:

- about 75% vegetable oil or animal fat,

- about 22% alcohol (such as methanol)

- a catalyst (lye, for instance)

This process being done, they are mingle meticulously for about an hour; after that, they are left to patch up. The glycerin drops to the bottom and biodiesel rises on top.

There are three fundamental techniques of biodiesel production from vegetable oils and animal fats:

• transesterfication

• the direct esterfication of the catalyzed acid of the oil with methanol

• the transformation of the oil acid into fatty acids, and then the conversion to methyl esters with acid catalysis

People's need to protect the environment and the perception upon biodiesel as an important ecological alternative to the all consuming traditional energy sources generated the birth of what may be called biodiesel industry. Over the last decades, it has accomplished a considerable growth, due to a significant boost in general production and production volume and an evolution from substantial lot plants to larger-scale permanent producers.

BIODIESEL

What is Biodiesel?

Biodiesel is a clean-burning sustainable diesel fuel substitute made from vegetable oil and animal fats. The oil used to make biodiesel comes from the farm, not from oil wells. Since plants produce biodiesel by absorbing carbon dioxide from the air, it is a carbon-neutral fuel.

Diesel engines today only remotely resemble the engine that Rudolf Diesel designed. His engine was initially designed to run on coal dust. Moreover, for the first few decades, they could run on just about anything that would burn. Even as late as the 1970's, automotive diesel engines were extremely tolerant when it comes to the fuel they could use. Today's diesel engines are much more efficient and cleaner than the early versions, partly because they run on a narrowly defined diesel fuel. It allows the engine manufactures to tune the fuel system for improved efficiency. Biodiesel and Diesel fuel are very similar in their physical and chemical characteristics. They are similar enough that modern diesel engines can run on biodiesel.

You can blend Biodiesel with diesel fuel at any ratio. You find biodiesel blends like B20, which is 20% biodiesel and 80% diesel, at truck stops and gas stations. Blending biodiesel with diesel fuel results in a fuel that burns cleaner than diesel alone. Everyone has seen those black clouds, or soot, pouring from the exhaust of a diesel truck when it is under load. However, when we blend biodiesel with diesel, the soot is reduced, and dirty diesels can become clean diesels.

We make biodiesel from vegetable oil; biodiesel is NOT vegetable oil. We can add vegetable oil to diesel as a diesel fuel extender, but only for a few tanks. Blending vegetable oil with diesel over the long run damages the engine and shortens its usable life. Vegetable oil is thicker or more viscous than diesel and does not burn efficiently in a fuel system designed for diesel fuel. So, we chemically alter vegetable oil into biodiesel using a process called transesterification.

Vegetable oil is three long chains of hydrocarbons (carbon and hydrogen atoms) glued to an alcohol molecule called glycerin. The chemical process swaps out the glycerin molecule with three methanol molecules. Since the reaction is reversible, we use an excess of methanol to drive the reaction in the direction of methyl esters (biodiesel). Transesterification needs a strong base as a catalyst to get the reaction going. The glycerin separates from the biodiesel with the help of gravity.

Glycerin is a high-value product used in the medical and food industries. It has numerous uses from sweeteners to fillers and lubricants. Unfortunately, homebrewers make a glycerin that is not pure and not worth much.

This innovative fuel is becoming more widely available. It can be found nationwide at a few select places or purchased directly from producers and marketers. It is a bit more expensive than traditional fuels, but as the demand for safe, biodegradable fuel alternatives increases, the price of biodiesel should rapidly fall. The cheapest way to acquire biodiesel is to make it yourself at home.

Biodiesel is a diesel fuel substitute used in diesel engines made from renewable materials such as:

Plant oils: canola, camelina, soy, flax, jatropha, mahua, pongamia pinnata, mustard, coconut, palm, hemp and sunflower;

Waste cooking oil: yellow or tap grease;

Other oils: tall, fish, and algae;

Animal fats: beef or sheep tallow, pork lard, or poultry fat; and

Potentially from cellulosic feedstock consisting of agriculture and forest biomass.

Warranty

The majority of North American engine manufacturers now endorse up to a B5 biodiesel blend. The Engine Manufacturers Association issued a technical statement indicating biodiesel use up to B5 should not cause engine or fuel systems problems. As biodiesel is more widely tested and used, manufacturers will be in a better position to support the use of higher blends. Warranty coverage of B20 and higher is offered by select manufacturers under specific conditions.

However, similar to using regular diesel, some manufacturers may limit the scope of their warranties by stating that failures from the use of any fuel cannot be attributed to a factory defect. Therefore, the cost of repair under these circumstances (if any) would not be covered by certain warranties.

It is recommended that you contact your engine manufacturer, dealer, or consult your equipment owner's guide for more information on warranty coverage; particularly, if you plan to use blends above B5.

Operability

In general, the NRDDI projects increases the confidence in the use of up to B5 in engines under the conditions encountered, as there were no engine failures reported.

Cold Weather

Like petroleum diesel, biodiesel can form crystals in cold weather which can lead to filter plugging. Laboratory tests show that a biodiesel blend forms crystals at a higher temperature than petroleum diesel. Actual experience with cold weather operations is influenced by many factors including the type of feedstock used as some types of biodiesel form crystals at lower temperatures than others, depending on the feedstock and characteristics of the fuel.

Potential solutions to cold weather problems are similar to those for petroleum diesel. They include using fuel additives and engine block or fuel filter heaters and storing vehicles in a building.

Several studies including the NRDDI projects have shown the successful use of biodiesel blends in cold weather up to a certain low concentration. It is important for the fuel provider to choose the right biodiesel formulation, and the fuel blend level is adjusted, to meet the Canadian General Standards Board recommended temperature specifications for the season and region of use. If you are unfamiliar with this subject it is recommended you discuss cold weather use with your fuel provider.

Cleansing Effect

Biodiesel acts like a mild solvent and has a cleansing effect. It "cleans out" sediments formed over time in equipment and storage fuel tanks which can cause occasional filter plugging, especially in the early stage of switching from petroleum diesel to biodiesel blends. However, when it is used in low concentrations, such as B5, it should not cause major issues. The NRDDI projects did not encounter any engine failures due to the cleansing effect.

In its research, the US National Renewable Energy Lab (NREL) found the cleansing effect should not be an issue with B5 and lower blends. Although, it is still wise to keep some extra filters on hand and monitor potential filter clogging a little more closely at the beginning.

Material Compatibility and Older Equipment

Older diesel engines were not necessarily designed to use biodiesel blends. Certain parts such as seals, gaskets, and connectors made with non-compatible rubbers or metals can be altered if the biodiesel blend is high. Consulting your engine manufacturer or equipment owner's guide (for the specific model-year of the equipment) will help you assess what maximum blend level is recommended. Today, most engine manufacturers accept the use of blends up to B5 without voiding the warranty; however, the blend must always meet industry approved standards.

This following guide (PDF, 1.6 MB) identifies which materials are not compatible with biodiesel (p.23, p.50 and p.51).

NREL has found there have been no significant material compatibility issues with B20 (unless the B20 has been oxidized). NREL found B20 or lower blends minimize most issues associated with materials compatibility. This conclusion provides even more confidence in the minimal effects which can be expected with B5.

In the NRDDI projects, where vehicle model years varied from 1967 to 2010 but most were of model year 1994 or newer, operation on B5 was not shown to cause any significant loss-of service incidents.

It is recommended that you consult your dealer or vehicle guide; particularly, if you plan to use biodiesel blends above B5 in older equipment.

Long-term Effects

Since an increasing number of engine manufacturers are endorsing B5 the NRDDI projects did not generally examine long term effects, such as wear, materials compatibility or longevity and did not find any issues which could be specifically attributed to these effects.

Stationary and Heating Equipment

Although there is limited experience with the use of biodiesel as a heating fuel in Canada, no significant adverse effects from using biodiesel in low-level blends in furnaces and boilers have been reported from initial trials. The NRDDI Imperial Oil project (PDF, 318 KB, in English only) tested biodiesel blends in space-heating furnaces. The results found that biodiesel use in furnaces should not exceed B10 in order to be compatible with existing seals in fuel pumps for late-model equipment. It remains a good practice to contact your furnace manufacturer; particularly, if you plan to use blends above B10.

The use of B5 in power generators in remote northern Canadian locations was demonstrated in both warm and cold seasons in the NRDDI Manitoba Hydro project (PDF, 2.1 MB)without any issues or requiring additional maintenance.

Performance

Biodiesel-powered engines have been shown to deliver similar torque and horsepower as diesel-powered engines.

Fuel Economy

Neat biodiesel (B100) has about 8% less energy content than diesel. As the diesel blend level is lowered, differences in energy content become proportionally less significant; blends of B5 or lower cause no easily noticeable differences in fuel consumption in comparison to diesel.

It is important to keep in mind your driving habits – when and where you drive, how often, the speed you travel, your aggressiveness on the road and other factors – have a lot to do with your vehicle's fuel consumption and costs. Inefficient driving practices can cause a greater increase in your fuel consumption and costs in comparison to switching your fuel from petroleum diesel to a B5 biodiesel blend.

Storage

All fuels have a limited shelf life. Long-term storage and storage with temperature variations leads fuel to degrade. Although biodiesel has been known to have a shorter shelf life than most petroleum diesels, fuel standards are designed to ensure all fuels have adequate long term performance. Adding proper additives is expected to address long-term storage concerns.

As summer fuel may be less suitable for winter, it remains a good practice to deplete fuel supplies before the season changes to ensure fuel remains appropriate for the expected temperature.

For low-level blends of biodiesel such as B2 and B5, end-users should be able to keep using their existing tanks. However, as mentioned above it is advised to keep extra filters on hand due to biodiesel's cleansing effect.

Safety

B100 is considerably less flammable than petroleum diesel, presents low-pressure storage at ambient temperatures, and is safer to handle and transport.

Why Produce My Own Biodiesel?

With the rising gas prices of today many people have begun to look for an alternative way to put fuel in their cars. A biodiesel kit fits the bill for many people who have started making this oil saving fuel in their own backyards. And why not make your own fuel, the hit to wallets and bank accounts with high gasoline prices makes finding alternative fuel sources imperative these days, particularly with the price of gas and oil in the news on a daily basis.

A biodiesel kit is a great way to save a substantial amount of money on your fuel costs. The basic function of such a kit is to convert vegetable oil into biodiesel, a fuel source that can be used in vehicles that currently run on diesel fuel. It is important to note that not all car companies endorse the use of biodiesel fuel in their vehicles and will void the warranty if an engine is damaged by its use. Be sure to research this before buying a new vehicle with a diesel engine. For older diesel cars and truck there are engine conversion kits that will make the change to cleaner burning biodiesel much easier.

There is a minimal amount of supplies needed to use a biodiesel kit. You will need basic tap water, methanol and access to a good amount of vegetable oil, which can usually be obtained from restaurants and such. You can set the kit up just about anywhere, although most people do put them outside in a shed or under an overhang. You will be making a fuel oil so setting up in the house may not be a real good idea.

When it comes to pricing a kit the more features to be found the more expensive the kit. The most expensive models are built to prevent fumes and odors from escaping during the refining process. Having an odorless refining process may be important for those wishing to refine their fuel near their home. If you live on a farm or a large acreage this may not be as important as the kit can be located farther from the house. Of course if you can afford it go ahead and get the more expensive closed system just because they are very easy to use.

If for some reason you already have a large amount of cooking oil that you need to dispose of then a biodiesel kit would work perfectly. It does take more time to process used oil but the results are the same, a more environmentally friendly fuel that can be put into a diesel powered vehicle.

Considering that the price of gasoline is on an ever rising trend making biodiesel for personal use can be seen as a smart move. Having your own biodiesel kit producing a constant supply of a low cost environmentally friendly fuel right in your own backyard would remove the sting of high pump prices.

Processor Can Save You Money

Biodiesel is an alternative fuel that is made from renewable resources, burns cleanly, is safe and can be used in any diesel engine vehicle. Biodiesel is being considered the fuel of the future. It has many benefits, but perhaps the biggest is that you can make it in your own home.

You can find a biodiesel processor for use at your home through an online search. There are many different manufacturers that will sell and ship to you a biodiesel processor. With the biodiesel processor you can turn regular vegetable oil into a fuel you can use instead of costly traditional fuels.

The process of making biodiesel involves four steps. These four steps are combined with the use of a biodiesel processor, making the whole process easy. The biodiesel processor will mix the ingredients, store them and even heat the oil. It will also keep everything contained during the separating process and wash the fuel so it is ready for use when the processor is finished. Additionally, the biodiesel processor can store the finished fuel so it is ready whenever you need it.

Biodiesel is a very safe fuel. It is actually safer than your average table salt. The process of making biodiesel is also safe because biodiesel does not involve any harmful ingredients and it does not produce harmful byproducts. However, you still need to keep safety in mind when working with a biodiesel processor because the process does involve heating up oil to high temperatures. Being careful is important to avoid burns.

You should look at your biodiesel processor as an investment. It is going to be rather pricey, usually in the thousands of dollars. However, when you compare that a gallon of biodiesel will cost you 70 cents to the cost of an average gallon of gas, then you will see why this is a wise investment. Making and using your own fuel helps give you control over your money and lets you get rid of the burden of high gas prices.

A biodiesel processor is the key to using one of the safest, cleanest and cheapest fuels available. Imagine being able to make your own biodiesel fuel whenever you need it. Imagine never having to run to the gas station and watch your hard earned dollars slip away. You can with a biodiesel processor. All you need are some simple ingredients and you are well on your way to doing your part in making this country clean and free from the confines of foreign oil.

Where can I buy biodiesel?

Being aware of environmental problems and issues in getting petroleum products, consumers are looking for other forms of fueling their engines. And lucky for owners of diesel powered engines there are a number of alternative fuels available today. One of which is bio diesel.

The main question clients are thinking of is where to buy biodiesel fuel. Many of the gas stations now are providing bio diesel as part of their products. S if you have a diesel engine and you have a filling station offering bio diesel fuel then it would be a breeze for you to use bio diesel.

In the United States, almost all of the states would have filling stations that have bio diesel available. Aside from big oil stations a lot of small businessmen have established their own filling station of pure bio diesel. And the main advantage is that diesel engines readily accept bio diesel.

Bio diesel can come in pure form containing products from developed waste veggie oil. Where to buy biodiesel fuel such as these can pretty well be in your backyard or in your own neighborhood. Oil companies are also mixing bio diesel with normal diesel. These are also named bio diesel.

These bio diesel mixes are labeled B20, B100, B99. Basically it would describe how much pure bi diesel is included in the mixture. B20 would have 20% bio diesel and 80% petroleum diesel. B100 is the pure form of bio diesel. B99 is commonly used because the 1% petroleum diesel included prevents mold build up.

Since the demand for bio diesel is at the rise, there are numerous pumps providing this safe gas. And in the future, it is predicted that this would be as common as petroleum gas. So if you were to ask where to buy biodiesel fuel? Then the answer is everywhere.

Why is not sold everywhere?

BIOFUELS HAVE BEEN around as long as cars have. At the start of the 20th century, Henry Ford planned to fuel his Model Ts with ethanol, and early diesel engines were shown to run on peanut oil.

But discoveries of huge petroleum deposits kept gasoline and diesel cheap for decades, and biofuels were largely forgotten. However, with the recent rise in oil prices, along with growing concern about global warming caused by carbon dioxide emissions, biofuels have been regaining popularity.

Gasoline and diesel are actually ancient biofuels. But they are known as fossil fuels because they are made from decomposed plants and animals that have been buried in the ground for millions of years. Biofuels are similar, except that they're made from plants grown today.

Much of the gasoline in the United States is blended with a biofuel ethanol. This is the same stuff as in alcoholic drinks, except that it's made from corn that has been heavily processed. There are various ways of making biofuels, but they generally use chemical reactions, fermentation, and heat to break down the starches, sugars, and other molecules in plants. The leftover products are then refined to produce a fuel that cars can use.

Countries around the world are using various kinds of biofuels. For decades, Brazil has turned sugarcane into ethanol, and some cars there can run on pure ethanol rather than as additive to fossil fuels. And biodiesel a diesel-like fuel commonly made from palm oil is generally available in Europe.

On the face of it, biofuels look like a great solution. Cars are a major source of atmospheric carbon dioxide, the main greenhouse gas that causes global warming. But since plants absorb carbon dioxide as they grow, crops grown for biofuels should suck up about as much carbon dioxide as comes out of the tailpipes of cars that burn these fuels. And unlike underground oil reserves, biofuels are a renewable resource since we can always grow more crops to turn into fuel.

Using biofuels as jetfuel also offers a solution to carbon emissions from air travel. In 2016, United

Airlines announced a new iniative to integrate biofuel into its energy supply with the hopes of reducing greenhouse gas emissions by 60 percent. Because commercial air travel comprises a significant amount of all carbon dioxide emissions, airlines and environmental advocates readily seek alternative fuel sources.

Downsides and benefits of using biodiesel

Unfortunately, it's not so simple. The process of growing the crops, making fertilizers and pesticides, and processing the plants into fuel consumes a lot of energy. It's so much energy that there is debate about whether ethanol from corn actually provides more energy than is required to grow and process it. Also, because much of the energy used in production comes from coal and natural gas, biofuels don't replace as much oil as they use.

Biofuels have also become a point of contention for conservation groups that argue crops would go to better use as a source of food rather than fuel. Famed primatologist Jane Goodall has gone so far as to warn that harvesting sugarcane and oil palms for biofuels would have devastating effects on rainforests.

For the future, many think a better way of making biofuels will be from grasses and saplings, which contain more cellulose. Cellulose is the tough material that makes up plants' cell walls, and most of the weight of a plant is cellulose. If cellulose can be turned into biofuel, it could be more efficient than current biofuels, and emit less carbon dioxide.

Advantages And Disadvantages Of Biodiesel Fuel

Compared to other alternative fuels, biodiesel fuel supports some unique features and qualities. Unlike any other alternative fuels, it has successfully passed all the health effects testing requirements, meeting the standards of the 1990 Clean Air Act Amendments.

Advantages of biodiesel fuel

• Biodiesel fuel is a renewable energy source unlike petroleum-based diesel.

• An excessive production of soybeans in the world makes it an economic way to utilize this surplus for manufacturing the Biodiesel fuel.

• One of the main biodiesel fuel advantages is that it is less polluting than petroleum diesel.

• The lack of sulfur in 100% biodiesel extends the life of catalytic converters.

- Another of the advantages of biodiesel fuel is that it can also be blended with other energy resources and oil.

- Biodiesel fuel can also be used in existing oil heating systems and diesel engines without making any alterations.

- It can also be distributed through existing diesel fuel pumps, which is another biodiesel fuel advantage over other alternative fuels.

- The lubricating property of the biodiesel may lengthen the lifetime of engines.

Disadvantages of biodiesel fuel

- At present, Biodiesel fuel is bout one and a half times more expensive than petroleum diesel fuel.

- It re uires energy to produce biodiesel fuel from soy crops, plaus there is the energy of sowing, fertilizing and harvesting.

- Another biodiesel fuel disadvantage is that it can harm rubber hoses in some engines.

- As Biodiesel cleans the dirt from the engine, this dirt can then get collected in the fuel filter, thus clogging it. So, filters have to be changed after the first several hours of biodiesel use.

- Biodiesel fuel distribution infrastructure needs improvement, which is another of the biodiesel fuel disadvantages.

We hope you found the above article on biodiesel fuel advantages and disadvantages both informative and useful.

BIODIESEL BENEFITS

The smartest technologies deliver benefits to multiple interests, including improved economy, and a positive impact on the environment and governmental policies.

The role of the biodiesel industry is not to replace petroleum diesel, but to help create a balanced energy policy with the most benefit to the United States. Biodiesel is one of several alternative fuels designed to extend the usefulness of petroleum, and the longevity and cleanliness of diesel engines.

The ultimate goal is to contribute to building a stronger, more self-sufficient community by way of a community-based biodiesel production model. A community-based biodiesel distribution program benefits local economies, from the farmers growing the feedstock to local businesses producing and distributing the fuel to the end consumer. The money stays in the community while reducing impact on the local environment and increasing energy security.

Biodiesel has many environmentally beneficial properties. The main benefit of biodiesel is that it can be described as 'carbon neutral'. This means that the fuel produces no net output of carbon in the form of carbon dioxide (CO_2). This effect occurs because when the oil crop grows it absorbs the same amount of CO_2 as is released when the fuel is combusted. In fact this is not completely accurate as CO_2 is released during the production of the fertilizer required to fertilize the fields in which the oil crops are grown. Fertilizer production is not the only source of pollution associated with the production of biodiesel, other sources include the etherification process, the solvent extraction of the oil, refining, drying and transporting. All these processes require an energy input either in the form of electricity or from a fuel, both of which will generally result in the release of green house gases. To properly assess the impact of all these sources requires use of a technique called life cycle analysis. Our section on LCA looks closer at this analysis. Biodiesel is rapidly biodegradable and completely non-toxic, meaning spillages represent far less of a risk than fossil diesel spillages. Biodiesel has a higher flash point than fossil diesel and so is safer in the event of a crash.

Depth information on the benefits of biodiesel

Easy To UseNo vehicle modifications or special fueling equipment just pump and go.

Power, Performance and Economy Proven performance and economy make biodiesel a renewable winner.

Emissions & Greenhouse Gas Reduction With lower exhaust emissions biodiesel is helping to reduce pollution and improve health. Lower CO_2 emissions help reduce the impacts of global warming.

Energy Balance & Security Biodiesel production and use at home, biodiesel helps reduce the need for foreign oil.

Toxicity, Biodegradability, Safety & Recycling Less toxic than table salt, biodiesel has minimal environmental impact. With a high flash point, biodiesel is safer to handle and store than petroleum diesel. When made from used oils and fats, biodiesel helps ensure proper recycling of former waste products.

Economic Development Biodiesel helps communities by keeping energy dollars at home.

Easy to Use

One of the great advantages of biodiesel is that it can be used in existing engines, vehicles and infrastructure with practically no changes. Biodiesel can be pumped, stored and burned just like petroleum diesel fuel, and can be used pure, or in blends with petroleum diesel fuel in any proportion. Power and fuel economy using biodiesel is practically identical to petroleum diesel fuel, and year round operation can be achieved by blending with diesel fuel.

Engine and Vehicles

All diesel engines and vehicles can use biodiesel or biodiesel blends. Certain older vehicles built before 1993 may require replacement of fuel lines which contain natural rubber, as biodiesel can cause these lines to swell or crack.

Blending and Switching with Diesel Fuel

Biodiesel can be used 100% (B100) or in blends with petroleum diesel fuel. Blends are indicated by B##, which correspond to the percentage of biodiesel in the blended fuel. For example, a 20% blend of biodiesel with 80% diesel fuel is called B20. When biodiesel is first used in a vehicle, it may release fuel tank deposits which can lead to fuel filter plugging. After this initial period, a user can switch between biodiesel and petroleum diesel whenever needed or desired, without modification.

Power, Performance and Economy

Many alternative fuels have difficulty gaining acceptance because they do not provide similar performance to their petroleum counterparts. Pure biodiesel and biodiesel blended with petroleum diesel fuel provide very similar horsepower, torque, and fuel mileage compared to petroleum diesel fuel. In its pure form, typical biodiesel will have an energy content 5%-10% lower than typical petroleum diesel. However it should be noted that petroleum diesel fuel energy content can vary as much as 15% from one supplier to the next. The lower energy content of biodiesel translates into slightly reduced performance when biodiesel is used in 100% form, although users typically report little noticeable change in mileage or performance. When blended with petroleum diesel at B20 levels, there is less than 2% change in fuel energy content, with users typically reporting no noticeable change in mileage or economy.

Superior Lubrication for Your Engine

The injection system of many diesel engines relies on the fuel to lubricate its parts. The degree to which fuel provides proper lubrication is its lubricity. Low lubricity petroleum diesel fuel can cause premature failure of injection system components and decreased performance. Biodiesel provides excellent lubricity to the fuel injection system. Recently, with the introduction of low sulfur and ultra low sulfur diesel fuel, many of the compounds which previously provided lubricating properties to petrodiesel fuel have been removed. By blending biodiesel in amounts as little as 5%, the lubricity of ultra low sulfur diesel can be dramatically improved, and the life of an engine's fuel injection system extended.

Biodiesel in Cold Weather

Just like petroleum diesel fuel, biodiesel can gel in cold weather. The best way to use biodiesel during the colder months is to blend it with winterized diesel fuel.

Emissions

Biodiesel is the only alternative fuel to successfully complete the EPA's rigorous emissions and health effects study under the Clean Air Act. Biodiesel provides significantly reduced emissions of carbon monoxide, particulate matter, unburned hydrocarbons, and sulfates compared to petroleum diesel fuel. Additionally, biodiesel reduces emissions of carcinogenic compounds by as much as 85% compared with petrodiesel. When blended with petroleum diesel fuel, these emissions reductions are generally directly proportional to the amount of biodiesel in the blend.

Is it sustainable?

A Safe and Stable Fuel

Biodiesel is safer to handle than petroleum fuel because of its low volatility. Due to the high energy content of all liquid fuels, there is a danger of accidental ignition when the fuel is being stored, transported, or transferred. The possibility of having an accidental ignition is related in part to the temperature at which the fuel will create enough vapors to ignite, known as the flash point temperature. The lower the flash point of a fuel is, the lower the temperature at which the fuel can form a combustible mixture. For example, gasoline has a flash point of -40 F, which means that gasoline can form a combustible mixture at temperatures as low as -40 F. Biodiesel on the other hand has a flash point of over 266 F, meaning it cannot form a combustible mixture until it is heated well above the boiling point of water. It is rare that fuel is subjected to these types of conditions, making biodiesel significantly safer to store, handle, and transport than petroleum diesel. In fact, the National Fire Protection Association classifies biodiesel as a non-flammable liquid.

Recovering Energy Resources

Biodiesel can be made from many different oils and fats, including many waste products. Waste cooking oil, normally disposed of or used in animal feed mixtures can be converted to high quality biodiesel using a process employed by companies such as Pacific Biodiesel Technologies. The use of used cooking oils as a biodiesel feedstock has increased their value significantly in recent years, making proper collection and recycling of these oils more cost effective, and lowering the volume of these oils destined for sewers and landfills. Other low value oils and fats which can be made into biodiesel include yellow grease, inedible tallow, and trap grease. In one example of the benefits of how biodiesel production can increase recycling, the Pacific Biodiesel production facilities in the Hawaiian islands have diverted nearly 190,000 tons of used cooking oil and grease trap waste since they began production.

Energy Dollars Stay In Communities

Since biodiesel is a fuel which can be created from locally available resources, it's production and use can provide a host of economic benefits for local communities. The community-based model of biodiesel production is particularly beneficial. In this model, locally available feedstocks are collected, converted to biodiesel, then distributed and used within the community. This model keeps energy dollars in the community instead of sending them to foreign oil producers and refineries outside the community. The peripheral benefits of this type of model are different for each case, but can include:

Increased tax base from biodiesel production operations.

Jobs created for feedstock farming and/or collection.

Skilled jobs created for biodiesel production and distribution.

Income for local feedstock producers and refiners.

Sustainable Farming and Value Added Agriculture

Biodiesel feedstock can come from a variety of agricultural crops. When these crops are grown in a sustainable manner, using good stewardship practices, there are long term benefits to farmers, farming communities and the land. Many crops which yield oils used for biodiesel production can be a beneficial rotation for other food crops, including soybeans when used in a traditional corn rotation, and canola when used in a wheat rotation. Using crops in rotation can improve soil health and reduce erosion. The overall impacts of growing energy crops are complex, with thousands of variables. However, the added value created for oilseed crops by the production of biodiesel is a tangible benefit for farming communities, and when coupled with sustainable farming practices can provide benefits to farming communities and the environment.

Sustainable Biodiesel Production

Since there are multiple feeds tocks from which to make biodiesel, plant operators can opt for the least expensive feedstock currently available, if they have a multiple-feedstock system. This flexibility makes producers less subject to price fluctuations.

CHAPTER 2

HOW DOES A DIESEL ENGINE WORKS

What is a diesel engine?

Like a gasoline engine, a diesel engine is a type of internal combustion engine. Combustion is another word for burning, and internal means inside, so an internal combustion engine is simply one where the fuel is burned inside the main part of the engine (the cylinders) where power is produced. That's very different from an external combustion engine such as those used by old-fashioned steam locomotives. In a steam engine, there's a big fire at one end of a boiler that heats water to make steam. The steam flows down long tubes to a cylinder at the opposite end of the boiler where it pushes a piston back and forth to move the wheels.

How Does a Diesel Engine Work?

You have just bought that stunning used 1986 Vista sundeck trawler and she is everything you needed in a yacht. The good sundeck, the master stateroom and the big fly bridge are just right. She also has twin Volvo diesels and a Onan gen set too. But you've been thinking, how do diesel engines function? You have never owned a diesel before.

In theory, diesels and gasoline engines are rather similar. They are both internal combustion motors designed to change the chemical energy available in diesel or gasoline into mechanical energy. This mechanical energy moves pistons up and down inside piston chambers. The pistons are connected to a crankshaft, and the up-and-down motion of the pistons, recognized as linear motion, creates the rotary motion needed to rotate the prop on your used trawler or motor yacht.

Today, where diesel or gasoline prices are increasing as a aftermath of spiraling demand and decreasing supply, we need to choose a cost effective fuel to power our vessels. After the invention of the diesel motor in 1892 by Rudolph Diesel in Augsburg Germany, the diesel engine has proved to be extremely efficient and cost effective. In1894 Rudolph Diesel was nearly killed when his prototype engine blew up. But that explosion established that diesel can be ignited without a spark plug.

A diesel motor is truly a bio-fuel engine. Diesel's first engine ran on peanut oil. In practice, a diesel can run on peanut oil, vegetable oils, synthetic oils, and even hydraulic fluids. Rudolf Diesel even experimented operating earlier diesels with gun powder. Handling and storing of the gun powder soon laid to rest that idea.

After oil was discovered to be a easily available resource, a fuel we now call diesel fuel was processed to power diesel engines. Diesel fuel is priced somewhat higher than gas but diesel has a higher energy density, i.e. more power can be withdrawn from diesel as compared with the same amount of gasoline. Therefore, diesel engines provide greater power, making it an unmistakable choice for big used trawlers and motor yachts. Diesel is heavier and oilier compared with gasoline. Diesel fuel has a high flash point making storage aboard a boat very safe.

The easiest way to remember how a diesel engine works is by remembering the phrase "suck, squeeze, bang, and blow".

This relates to a cycle of 4 strokes recognized as the OTTO cycle.

First of all, air is drawn into the cylinder (suck). The air is then compressed by the movement of the piston, and fuel is injected as a vapor just before the piston contacts the top of the cylinder (squeeze). The compression raises the temperature of the air; which causes the diesel fuel to combust (bang). Eventually, the waste gases are blown from the piston chamber (blow) and into the exhaust system.

A diesel injection pump is responsible for injecting fuel into the firing cylinders of diesel engines. It is important to remember that, different from gasoline-powered engines, spark plugs are not used to ignite the fuel. They rely entirely on the compressing of the diesel in the piston chamber to result in combustion. Therefore, diesel injection pumps are extremely important and must be built tough to create the compression values of up to 15,000 psi necessary for the engine's operation.

Naturally aspirated diesel engines just pull in the air (suck) to begin the combustion cycle. These diesels produce less power than their turbocharged cousins. Turbocharging is the mechanical forcing of air into the engine permitting it to produce more power.

Turbocharged diesel engines refer to any diesel engine with a turbocharger. Turbo charging is the norm rather than the exception in bigger and faster motor yachts. As with any turbocharged engine, turbo diesels can offer greater power outputs, lower emissions levels, improved efficiency than their naturally aspirated counterparts.

Hi power engines re uires stronger (and thus heavier) internal component parts such as the pistons and crankshaft to withstand the continuous hammering from the diesel engine's operating cycle. Therefore, the design of a diesel engine is made to take hundreds of hours of constant use under load. I am told by the Westerbeke representative of one engine, still in use today that has 30,000 hours on her, and she is still operating fine.

Diesel engines can sustain damage as a result of misapplication or misuse - principally internal glazing and carbon buildup. This is often found in generators caused by failure to run the engines not under a load - ideally diesel engines should run at least about 75% of their maximum rated load and Revolutions Per Minute. Short-term times of low load running are permissible providing the diesel engine is brought up to full load, or close to full load on a regular basis

How Does Bio-Diesel Work?

With ever increasing fuel costs, and concerns about greenhouse gas emissions damaging the climate, many of us are looking for alternatives. One of these is bio-diesel, but how does bio-diesel work and how well does it answer these problems? The quick answer is this fuel works in exactly the same way as regular diesel, and it can be used in most modern engines without any modification. Below are some points you should consider if you plan to make the switch.

Environment:

Making use of bio-diesel reduces greenhouse gas pollutants a lot more successfully compared with ethanol, gasoline, and standard diesel fuel. It will in addition help you to alter your power usage towards better renewable options. There's a 90% decrease in emissions if you choose to employ 100 % pure bio-fuel above fossil fuel diesel. The reason for this is that oil-bearing crops like soybeans soak up co2 (carbon dioxide) out of the surroundings as they grow. When this vegetable oil is made into fuel exactly the same quantity of carbon dioxide will be introduced into the environment when it is burned by your engine. This action is known as being "Carbon Neutral," since there's an identical trade in co2 usage by the plant and discharge by your engine. Burning non-renewable fuels continuously increases the co2 load of the surroundings, because the co2 in the fossil oil was absorbed by plants millions of years in the past.

Switching:

If you decide to make the switch to this fuel then you don't necessarily have to go from regular to full bio in one step. You can buy a bio and regular diesel blend, if you live in Europe chances are you already use 20% bio blend as commercially available fuels are required to contain at least this much in their composition. Whilst most modern engines can use the fuel in its pure form without modification, older engines may need to be converted. This is because older engines tend to have natural rubber seals, and bio-diesel degrades rubber over time. Modern engines with their synthetic silicone rubber components are not affected.

Better running:

Many people using bio-diesel report that their vehicles start better, and travel much more smoothly. So the advantages of this fuel definitely outnumber the disadvantages. This kind of technology can potentially be applied to numerous kinds of motor vehicles and applications: trucks, boats, buses, trains, vans, construction machinery, agricultural machinery, generators, and automobiles, in fact anything powered by regular polluting diesel can be almost instantly made greener just by switching to this fuel.

An additional favorable effect of this renewable fuel will be the improvement in the lifespan of the vehicle's power-plant. This is a significantly cleaner fuel, so the harmful effect of deposits of soot and sulfur on the motor is drastically decreased, plus it has a much better natural lubrication effect which reduces wear. All this means you will get a much longer life out of your car or truck.

As you can see if you are wondering how does bio-diesel Work? It works exactly like regular diesel, with the added advantages of being kinder to your engine and to the environment. There are many reasons to run your vehicle on clean renewable bio fuel, the advantages are obvious so what is stopping you from making the switch today?

How is a diesel engine different from a gasoline engine?

Gasoline engines and diesel engines both work by internal combustion, but in slightly different ways. In a gasoline engine, fuel and air is injected into small metal cylinders. A piston compresses (s□ueezes) the mixture, making it explosive, and a small electric spark from a sparking plug sets fire to it. That makes the mixture explode generating power that pushes the piston down the cylinder and (through the crankshaft and gears) turns the wheels. You can read more about this and watch a simple animation of how it works in our article on car engines.

Diesel engines are similar, but simpler. First, air is allowed into the cylinder and the piston compresses it but much more than in a gasoline engine. In a gasoline engine, the fuel-air mixture is compressed to about a tenth of its original volume. But in a diesel engine, the air is compressed by anything from 14 to 25 times. If you've ever pumped up a bicycle tire, you'll have felt the pump getting hotter in your hands the longer you used it. That's because compressing a gas generates heat. Imagine, then, how much heat is generated by forcing air into 14–25 times less space than it normally takes up. So much heat, as it happens, that the air gets really hot usually at least 500°C (1000°F) and sometimes very much hotter. Once the air is compressed, a mist of fuel is sprayed into the cylinder typically (in a modern engine) by an electronic fuel-injection system, which works a bit like a sophisticated aerosol can. (The amount of fuel injected varies, depending on how much power the driver wants the engine to produce.) The air is so hot that the fuel instantly ignites and explodes without any need for a spark plug. This controlled explosion makes the piston push back out of the cylinder, producing the power that drives the vehicle or machine in which the engine is mounted. When the piston goes back into the cylinder, the exhaust gases are pushed out through an exhaust valve and, the process repeats itself hundreds or thousands of times a minute!

How a four-stroke diesel engine works.

Four-stroke engines

Like a gasoline engine, a diesel engine usually operates by repeating a cycle of four stages or strokes, during which the piston moves up and down twice (the crankshaft rotates twice in other words) during the cycle.

Intake: Air (light blue) is drawn into the cylinder through the open green air inlet valve on the right as the piston moves down.

Compression: The inlet valve closes, the piston moves up, and compresses the air mixture, heating it up. Fuel (dark blue) is injected into the hot gas through the central fuel injection valve and spontaneously ignites. Unlike with a gas engine, no sparking plug is needed to make this happen.

Power: As the air-fuel mixture ignites and burns, it pushes the piston down, driving the crankshaft (red wheel at bottom) that sends power to the wheels.

Exhaust: The green outlet valve on the left opens to let out the exhaust gases, pushed out by the returning piston.

Two-stroke engines

In a two-stroke diesel, the complete cycle happens as the piston moves up and down just once. Confusingly, there are really three stages in a two-stroke cycle:

Exhaust and intake: Fresh air is blown into the side of the cylinder, pushing the old exhaust out through valves at the top.

Compression: The inlet and exhaust valves close. The piston moves up, compresses the air, and heats it up. When the piston reaches the top of the cylinder, fuel is injected and spontaneously ignites.

Power: As the air-fuel mixture ignites, it pushes the piston down, driving the crankshaft that sends power to the wheels.

Two-stroke engines are smaller and lighter than four-stroke ones, and tend to be more efficient since they produce power once during each rotation (instead of once during every two rotations, as in a four-stroke engine). This means they need more cooling and lubrication and suffer higher wear and tear.

What makes a diesel engine more efficient?

Diesel engines are up to twice as efficient as gasoline engines around 40 percent efficient, that is. In simple terms, that means you can go much further on the same amount of fuel (or get more miles for your money). There are several reasons for this. First, they compress more and operate at higher temperatures. A fundamental theory of how heat engines work, known as Carnot's rule, tells us that the efficiency of an engine depends on the high and low temperatures between which it operates. A Diesel engine that cycle through a bigger temperature difference (a higher hottest temperature or a lowest colder temperature) is more efficient. Second, the lack of a sparking-plug ignition system makes for a simpler design that can easily compress the fuel much more and compressing the fuel more makes it burn more completely with the air in the cylinder, releasing more energy. There's another efficiency saving too. In a gasoline engine that's not working at full power, you need to supply more fuel (or less air) to the cylinder to keep it working; diesel engines don't have that problem so they need less fuel when they're working at lower power. Another important factor is that diesel fuel carries slightly more energy per gallon than gasoline because the molecules it's made from have more energy locking their atoms together (in other words, diesel has a higher energy density than gasoline). Diesel is also a better lubricant than gasoline so a diesel engine will naturally run with less friction.

How is diesel fuel different?

Diesel and gasoline are quite different. You'll know this much if you've ever heard the horror stories of people who've filled up their car or truck with the wrong sort of fuel! Essentially, diesel is a lower-grade, less-refined product of petroleum made from heavier hydrocarbons (molecules built from more carbon and hydrogen atoms). Crude diesel engines that lack sophisticated fuel injection systems can, in theory, run on almost any hydrocarbon fuel hence the popularity of biodiesel (a type of biofuel made from, among other things, waste vegetable oil). The inventor of the diesel engine, Rudolf Diesel, successfully ran his early engines on peanut oil and thought his engine would do people a favor by freeing them from a dependency on fuels like coal and gasoline. If only he'd known!

Advantages and disadvantages of diesel engines

Diesels are the most versatile fuel-burning engines in common use today, found in everything from trains and cranes to bulldozers and submarines. Compared to gasoline engines, they're simpler, more efficient, and more economical. They're also safer, because diesel fuel is less volatile and its vapor less explosive than gasoline. Unlike gasoline engines, they're particularly good for moving large loads at low speeds, so they're ideal for use in freight-hauling ships, trucks, buses, and locomotives. Higher compression means the parts of a diesel engine have to withstand far greater stresses and strains than those in a gasoline engine. That's why diesel engines need to be stronger and heavier and why, for a long time, they were used only to power large vehicles and machines. While this may seem a drawback, it means diesel engines are typically more robust and last a lot longer than gasoline engines.

Pollution is one of the biggest drawbacks of diesel engines: they're noisy and they produce a lot of unburned soot particles, which are dirty and hazardous to health. In theory, diesels are more efficient, so they should use less fuel, produce fewer carbon dioxide (CO_2) emissions, and contribute less to global warming. In practice, there's some argument over whether that's really true. Some laboratory experiments have shown average diesel emissions are only slightly lower than those from gasoline engines, although manufacturers insist that if similar diesel and gasoline cars are compared, the diesels do indeed come out better. Other recent research suggests that even new diesel cars are highly polluting.

What about CO_2 emissions?

Diesel cars have contributed massively to reducing CO_2 emissions. Since 2002, buyers choosing diesel have saved almost 3 million tones of CO_2 from going into the atmosphere." Diesel engines do tend to cost more initially than gasoline engines, though their lower running costs and longer operating life generally offsets that.

There's no question that diesel engines will continue to power heavy vehicles, but their future in cars and lighter vehicles is becoming increasingly uncertain. The push toward electric cars has provided a powerful impetus to make gasoline engines lighter, more economical, and less polluting, and these improved gas engines undermine some of the perceived advantages of using diesels in cars. In the growing competition between affordable electric vehicles and improved gasoline cars, diesels may find themselves squeezed out altogether.

CHAPTER 3

COLLECTING MATERIALS

Here is a general list of biodiesel supplies you may need. It is important to keep in mind that high quality finished biodiesel can largely be dependent on the use of high quality biodiesel supplies. 'High quailty' should not to be confused with expensive supplies. High quality Biodiesel can be made inexpensively, but it is important to put some thought into the equipment you use.

Biodiesel Making Equipment:

- Biodiesel kits are a great way to be introduced to biodiesel. The most basic of kits will have you use straight vegetable oil and will have the catalyst and methanol already measured out for you simply heat and mix. Other kits give you enough materials to make several test batches, and the option of using WVO.

- If you plan on collecting waste vegetable oil for your stock, you will need 55-gallon drums, a pump, and also sampling containers.

- Filters are used for filtering WVO and even SVO before use. They can also be used for filtering the biodiesel after production. Photo courtesy of Iofaesofa

Some home brewers use blue jeans as a filter. Filters can be purchased as bag or cartridge. Some brewers don't filter at all – they just let it settle and pump off the top.

- Titration tests are used to calculate the amount of catalyst that will be needed to perform transesterification (needs will vary based on feedstock).

- Chemicals - for titration, catalysts, and dry wash resins (dry washing is optional).

- A scale to weigh the chemicals

- A mixer (not necessary) for dissolving the catalyst in the methanol, performing titration tests.

- A heater for heating the oil to the correct temperature for transesterification to take place, and some use a heater to dry the biodiesel.

- A reactor for transesterification to take place in.

- Settling tank where the mixture will set until the glycerin has settled to the bottom.

- Wet Wash or Dry Wash tank where the biodiesel is purified usually by water or ion exchange resins.

- A methanol recovery system for reclaiming remnant methanol for re-use (if desired).

- A biodiesel processor (kind of self explanatory) is a requirement for your biodiesel supplies! There are many options; they come in kits or fully assembled... or you can even build your own. Some are automatic; some must be periodically tended to.

- A pump for transferring oil to different containers during production.

- A storage tank for biodiesel may be made of aluminum, steel, teflon or fluorinated polyethylene. Many home brewers store theirs in a 55-gallon drum. Biodiesel should be stored in a clean, dry, dark environment. A storage tank for feedstock oil is also needed.

- SA storage Container for the chemicals - it must be air tight.

- Flow Meters to measure the Quantity of methanol and oil added to a batch.

- A centrifuge can be used for pre-treatment; to separate the water, oil, and solids; and is most commonly used as a speedy way to separate the glycerin from the biodiesel.

- A Rapid Biodiesel Tester or a Gas chromatograph (used by the more advanced producers) to check the Quality of the feedstock and level of free glycerin in the biodiesel.

• Particulate Test used for testing the particulate level, thus the quality of the finished biodiesel.

•Safety biodiesel supplies

to protect you skin and eyes from splashes from the harsh chemicals. Also safety equipment should be used around the methanol, or the fumes should be contained.

Biodiesel Safety: Biodiesel itself is a very safe substance. When compared to table salt, it is ten times less toxic. It biodegrades easily and has a much higher flashpoint than petroleum diesel.

Biodiesel safety glove

The lowest temperature at which the fuel will ignite when exposed to a flame is referred to as its flashpoint.

B100 must have a flashpoint greater than 266°F (130°C) to be ASTM certified. Flashpoint can vary due to different feed stocks used.

Methanol Biodiesel Safety Precautions:

Methanol is a highly flammable gas. Its boiling point is at 148.3F (64.6C). Its flash point in a closed container is 54F (12C), and its flash point in an open container is 60.1F (15.6C).

- Ingestion - as little as 1-4oz has caused fatalities. Immediate medical attention should be sought; the symptoms may be delayed for many hours. Vomiting should not be induced.

- Contact with skin – there are no known serious effects as long as there is not repeated exposure. If skin comes in contact with the gas, wash with soap and water for 15 minutes or more. If irritation occurs seek medical attention.

- Contact with eyes – flush eyes gently with water for a minimum of 15 minutes and seek medical attention.

- Inhalation – Relocate to fresh air and seek medical attention incase of delayed onset of a serious reaction. A person should be exposed to no more than 200ppm within an eight-hour period, and no more than 250ppm in a 15-minute period.

- Protection Measures - Safety goggles with side shields and appropriate chemical gloves are recommended. Use in a well ventilated area.

Lye and Potash Biodiesel Safety Precautions:

Lye and Potash are corrosive chemicals. They should not come in contact with aluminum, magnesium, brass, bronze, tin, chromium, or galvanized zinc because they can form hydrogen gas, which is highly explosive.

- Ingestion – may cause shock, nausea, diarrhea, vomiting, severe pain and permanent damage. Drink a glass of water if possible; do not induce vomiting. Medical attention should be sought immediately.

- Contact with skin – can cause burns and deep ulcers. Remove contaminated clothing and rinse skin for at least 15 minutes and seek medical attention.

- Contact with eyes – may cause burns and blindness. Flush eyes with water for 15 minutes and seek medical attention.

- Inhalation – may cause burns, coughing, difficulty breathing, and coma. Relocate to fresh air.

- Protection Measures – wear safety goggles with side shields and chemical resistant gloves. Use these chemicals in a well-ventilated area. Also, when dissolving lye in water in an open top container, wear a vapor-resistant mask.

For all chemicals involved in biodiesel production, chemical gloves and safety goggles should be worn. Respirators can be avoided by careful planning, such as using a sealed container for methanol recovery to take place in, or mixing your lye outdoors.

CHAPTER 4

SETTING UP THE WORKSHOP

Workshop planning and layout

A large number of beginners and hobbyists eventually become professionals, so the number of tools and machinery in the workshop is increasing, which leads to the need for the bigger workshop. Nowadays, not many people own enough large space to adapt it into a home workshop. Most people are setting their workshops in the basement, garage, or they are adapting some rooms that were originally intended for other purposes. For these reasons, planning and the proper workshop layout will enable you a good and pleasant work.

When planning the workshop layout, it is necessary to consider a number of factors that affect your work:

- What exactly do you want to do in the workshop: in order to make proper plan about required dimensions of the workshop and its organization, at first think about what are you going to use it for, what are your affinities, plans and working possibilities in the future;

- Estimate the costs and time required to equip the workshop: evaluate your options and what do you need for your workshop so it could become functional;

- You should plan the arrangement of the machines and worktables to ensure enough space to maneuver;

- The number of machines and workshop accessories (e.g. sawhorses, benches...);

- Calculate the number of workers / people in the workshop;

- Will you be making small objects such as toys and models, or you will manufacture large items such as doors and windows. This is very important when planning the space requirements, size of the machines, storage space for materials and finished products;

- space for storing lumber and other materials for processing, and the storage of finished products;

- Room for the finish: it is smart to plan a separate room, where you can coat your products with protective oils, paints and lacquer. It should be dust free, with good ventilation for removing various vapors during the drying of coatings, and it should have enough space for comfortable work, because each scratch or blur is noticeable on the coating, and that can seriously reduce aesthetic value of your work;

- How easy is the access to the home workshop: The workshop should have direct entrance from the outside (not from the house), because it will often be necessary to carry in or out some large, heavy or dirty things, and a lot of waste as well. It may require access for the freight vehicle. Besides that, various works that produce noise, dust or unpleasant smells will be often made in the workshop, so it's good that the workshop is not located in the immediate area of the buildings where the people live;

- Are the doors of appropriate size for all your needs, so that machinery, lumber and finished products could pass through them;

- To prevent shop theft with the security measures related to the workshop, but, in the case of the high value machinery and products in the workshop, it should also be considered the business insurance;

- Are the heating / air conditioning, isolation and ventilation adequate;

- Is the plumbing and consequently the drainage of waste water needed for your workshop;

- Are there enough electrical circuits to supply your power needs;

- Quality lighting;

- Dust removing system.

An ideal workshop does not exist. Each woodworker has its own unique interests and needs; a number of limiting factors affects the set-up of workshop. The most important thing is to create a workshop that will suit your needs and work habits in the best possible way and to equip your workshop with hand tools, power tools and machines that you need for the work.

Starting a Biodiesel Production Company – Sample Business Plan Template

1. Conduct Your Research and Feasibility Studies

Starting a biodiesel production company is not a business that can be started without adequate preparation. If indeed you want to be successful in any form business including starting your own biodiesel production company, then you must spend time and money on research and feasibility studies.

The truth is that if you effectively conduct your research and feasibility studies, you are able to make informed decision whether to go ahead with your desire of starting your own biodiesel production company or to call it quit when you know you don't have the capacity to confront the challenges that comes with starting and running such business.

2. Write a Business Plan

Your business plan is a document that clearly describes your business and outline goals and roadmaps that leads to the success of your business. It is indeed a document that describes the future or your business. Hence, if you truly want to start your own biodiesel production company, it will pay you to write a comprehensive business plan and ensure that you religiously follow the business plan as you run your business.

It is also advisable to hire professional business plan writers to help you write your business plan.

Although, you might have to pay them, but the truth is that it definitely would be worth your time and money.

3. Incorporate Your Biodiesel Company

If you have carried out your research and feasibility studies and you have also written your business plan, then the next thing for you to do is to incorporate your company. The truth is that you can't start a biodiesel production company without legally registering the business with the Federal Government of your country.

4. Apply for Operation License and Permit

Because of the nature and hazard involved in biodiesel production, you would be required to meet up with certain environmental protection standards before you can be issued license and permission to start your own biodiesel production company.

As a matter of fact, your facility would be inspected to ensure that you have in place spill prevention control and counter – measures plan, and also they would ensure that you meet up with other expected standard before your application for a license and permit can be approved and issued.

5. Lease or Acquire a Facility

Starting a biodiesel Production Company requires that you lease or acquire a facility that is outside residential areas because of the danger of air – pollution and hazardous discharge that goes into municipal waste system. It is only normal that your facility will first be inspected and approved before you can be allowed to go into the production of biodiesel. So, ensure that you get a facility that is located in an industrial area.

6. Employ Competent Staffs

A biodiesel production company is not a business that you can start as a sole proprietor; you would require competent employees to work with you. Therefore, when you are hiring employees, ensure that you only recruit people that know what it means to work in an industry that is as delicate as the biodiesel production company.

The truth is that serious caution should be taken while producing biodiesel because accidents in this industry can be very devastating and can even lead to death. So, you would not want to employ people that can't obey safety rules that are put in place by your company.

7. Market Your Biodiesel

You just have to formulate strategies that can help you effectively market your biodiesel. The fact that the use of biodiesel in the United States of America and in other countries of the world is on the increase doesn't mean that you won't have to take your time to study the market and know how to penetrate.

When you go out there to market your biodiesel, your selling point should be anchored on the fact that biodiesel is safer than petroleum diesel and that it has lower emission when compared to other fuels. It is important to state that the market for biodiesel is on the increase and if you position your company properly, you will have a large chunk of the available market.

It is no doubt biodiesel has been proven to be a good substitute for petroleum diesel and it is fact that the industry has a very bright future

Design and Construction of Biodiesel Production Plant

Transesterification is a technology based on chemical reaction of triglycerides with methanol to form methyl esters and glycerin in the presence of an alkaline catalyst. In principle, materials including edible, vegetable oils and animal fat can be transesterified. If the feedstock quality does not meet the specification, pre-treatment and/or esterification are required for the appropriate process.

Transesterification occurs in the mixing section of a two-stage mixer-settler unit, while the subsequent settling section allows the separation of methyl esters lighter and glycerin heavier. After rectification, the surplus methanol in the glycerin-water mixture can be reused in the process

Features

Multiple-feedstock and fully-automatic batch or continuous process

High product yield(~100%; 1 kg feedstock yields 1 kg biodiesel)

High quality biodiesel and glycerin

Good ROI

Minimizes operating and maintenance cost

Applications

Biodiesel- for all diesel engine machines and tools applicable

Glycerin- Pharmaceutical or industry applications

CHAPTER 5

BASICS TO UNDERSTAND THE CHEMICAL PROCESS

Derived from living or recently living organisms, or biomass, the basic composition of biofuels is more complex than the composition of fossil fuels. While fossil fuels consist only of carbon and hydrogen atoms, or hydrocarbons, biofuels contain oxygen atoms, and their chemical composition may include acids, alcohols and esters.

Biobutanol

Biobutanol is derived from biomass or produced by fermentation using organisms found in ruminant animals and soils. The basic composition of butanol consists of C (carbon), H (hydrogen) and O (oxygen). The chemical formula for the butanol molecule is $C_4H_{10}O$. Biobutanol provides more energy than ethanol and can be blended with gasoline to help reduce greenhouse gases. Any car that runs on gasoline can run on a biobutanol blend.

Biodiesel

Derived from vegetable oils and animal fats, biodiesel molecules are esters of long-chain fatty acids containing single chains of 12 to 24 carbon atoms. The esters contain an alcohol and a carboxylic acid. The carboxylic acid contains COOH (carboxyl), and the alcohol contains OH (hydroxide). Biodiesel burns cleaner than traditional diesel, producing less sulfur and fewer particulates. Biodiesel delivers slightly less energy than petroleum-based diesel, however, and is more corrosive to engine parts.

Ethanol

Derived from corn, sugar beets and sugar cane, other sources for ethanol production such as corn stover and switchgrass are under development. Containing carbon, hydrogen and a hydroxide group, the chemical formula for the ethanol molecule is C_2H_5OH. Any vehicle sold in the United States can run on E10, a blend of 10 percent ethanol and 90 percent unleaded gasoline. Providing about 50 percent of the energy of gasoline, combustion of ethanol is cleaner and produces less carbon monoxide but produces more smog.

Methanol

The simplest of alcohols, methanol can be derived from any plant material as well as landfill gas, power plant emissions and atmospheric carbon dioxide. The basic composition of methanol consists of carbon, hydrogen and hydroxide. The chemical formula for the ethanol molecule is CH3OH. Methanol combustion produces a lower volume of toxins than gasoline, fewer particulates and less smog. Methanol is less expensive than gasoline or ethanol, and the cost of modifying a vehicle to run on blends of methanol is low.

The Chemistry of Biofuels

Variety

Petroleum is a rather simple substance because almost all of the molecules are hydrocarbons. Though come in varying lengths and some are gases, others are liquid, and still others are solids, the fact that they are all made of hydrogen and carbon means that they all behave in roughly the same way. The same is not true of biofuel.

Biofuels contain oxygen. By adding just that one atom, the complexity of these molecules begins to rise dramatically. Whereas the major difference between fossil fuels revolves around whether they have single or double bonds and how long they are, the differences between biofuels are far more complex.

Because of the oxygen, biofuels can contain alcohols, esters, ethers, and acid groups. Each of these groups is a whole subset of organic chemistry and each has special reaction characteristics. Alcohols are really nothing like ester, which are not like acids at all. The net result of adding oxygen is a huge jump in complexity.

Biology and Synthesis

When dealing with fossil fuels, the processes are all chemical or physical. What is more, fossil fuels aren't really produced so much as they are refined. Petroleum already contains gasoline, diesel, and other useful compounds. All industry really does is refine the petroleum in order to separate out the parts. It is true that they do other things, like cracking, to improve yield, but the basic process of getting fuel from petroleum is relatively straightforward.

The same is not true of biofuels. Before a biofuel can ever be used, the feedstock first has to be produced. This requires a complex understanding of chemistry and biology. In many cases, organisms are genetically modified to improve yield and to reduce nutrient and water requirements. Thus, one must have a thorough grasp of biology before even planting a biofuel crop. After that, biofuel is either refined from oil or produced by algae and harvested. In either case, complex chemistry is involved to get useful fuels out of biological molecules.

ORGANIC CHEMISTRY BASICS

Organic chemistry is the study of the structure, properties, composition, reactions, and preparation of carbon-containing compounds, which include not only hydrocarbons but also compounds with any number of other elements, including hydrogen (most compounds contain at least one carbon-hydrogen bond), nitrogen, oxygen, halogens, phosphorus, silicon, and sulphur. This branch of organic chemistry was originally limited to compounds produced by living organisms but has been broadened to include human-made substances such as plastics. The range of application of organic compounds is enormous and also includes, but is not limited to, pharmaceuticals, petrochemicals, food, explosives, paints, and cosmetics. This article contains the organic chemistry basics.

Organic chemistry is a highly creative science in which chemists create new molecules and explore the properties of existing compounds. It is the most popular field of study for ACS chemists and Ph.D. chemists.

Organic compounds are all around us. They are central to the economic growth of the United States in the rubber, plastics, fuel, pharmaceutical, cosmetics, detergent, coatings, dyestuff, and agrichemical industries, to name a few. The very foundations of biochemistry, biotechnology, and medicine are built on organic compounds and their role in life processes. Many modern, high-tech materials are at least partially composed of organic compounds.

Organic chemists spend much of their time creating new compounds and developing better ways of synthesizing previously known compounds. This article contains the organic chemistry basics.

Organic Compounds (organic chemistry basics)

In addition to carbon and hydrogen, many organic compounds also include oxygen, nitrogen, chlorine, bromine or sulphur. Other elements are also possible but are less common. Organic compounds vary in complexity from very simple molecules of just a few atoms (e.g. methane, CH_4) to extremely long polymer chains that include many thousands of atoms forming each molecule. The huge number and variety of different organic compounds – with different physical and chemical properties – is due to the different ways in which (a relatively small number of) elements can link to each other, not due to the involvement of many different elements.

Carbon and its properties

Carbon compounds are present everywhere i.e. in the food that we eat, the clothes that we wear and even in the lead of the pencil by which we write. The atomic number of carbon is 6 and atomic mass is $12.01 gmol^{-1}$. Carbon is a member of the 14th group. According to the data, it is the seventeenth most abundant element found on earth. It is found in both free as well as in the combined state.

Carbon is found in both free as well as in the combined state. You can find it available as coal or graphite in the elemental state. Whereas it is present as metal carbonates, hydrocarbons, and carbon dioxide gas in the combined state. When it combines with other elements such as di-hydrogen, di-oxygen, chlorine and sulphur provides amazing arrays of materials that can vary from tissues to medicines.

In Organic Chemistry, everything is surrounded by carbon compounds. It is one of the essential components of the living organisms. There are two stable isotopes of carbon $12C$ and $13C$. After these two one more isotope of carbon is present $14C$. Carbon is used for radiocarbon dating and it is also a radioisotope with the half-life of 5770 years.

Properties of Carbon

Physical Properties of Carbon:

Carbon is a unique element. It occurs in many forms. Some of the examples of the pure form of carbon are coal and soot.

It is soft and dull grey or black in color.

One of the most important compounds of carbon is the charcoal, which is formed when carbon is heated in the absence in of air.

It occurs in a number of allotropic forms. Allotropes are nothing but forms of an element with varying physical as well as chemical properties.

The density of the different forms of carbon depends upon the origin of these elements. You will find some forms of carbon which are pure and some forms which are not pure like coal which is the mixture of both carbon and hydrogen.

Chemical Properties of Carbon:

Carbon compounds generally show 4 reactions, they are

Combustion reaction

Oxidation reaction,

Addition reactions

Substitution reaction.

As we all know that carbon in all forms needs oxygen, heat, and light and forms carbon dioxide. When it is burned in air to give carbon dioxide, it is called as combustion.

Let us get the concept of this using some examples when it is burnt in the air: When methane CH_4 is burnt in the presence of oxygen it gives us carbon dioxide, heat, and light.

Carbon compounds in fuels like diesel:

This is because when carbon or diesel is burnt in air, it produces a large amount of heat energy.

Combustion is of two types:

Complete Combustion

Incomplete Combustion

We all have seen cooking gas burning at home, they turn blue in color. Have you ever thought why it appears blue? The answer is that when a saturated hydrocarbon is burnt in the presence of oxygen then it gives blue color flame and this is because here fuel is burnt completely. In other words, you can understand it by knowing that the percentage of carbon present in saturated hydrocarbons is low and it gets easily oxidized.

Functional Groups

Bromine reacts with 2-butene to form 2, 3-dibromobutane.

It also reacts with 3-methyl-2-pentene to form 2,3-dibromopentane.

Instead of trying to memorize both equations, we can build a general rule that bromine reacts with compounds that contain a C=C double bond to give the product expected in addition to the double bond. This approach to understanding the chemistry of organic compounds presumes that certain atoms or groups of atoms known as functional groups give these compounds their characteristic properties.

Functional groups focus attention on the important aspects of the structure of a molecule. We don't have to worry about the differences between the structures of 1-butene and 2-methyl-2-hexene, for example, when these compounds react with hydrogen bromide. We can focus on the fact that both compounds are alkenes that add HBr across the C=C double bond in the direction predicted by Markovnikov's rule.

Functional Groups That Contain a CarbonylThe C=O group plays a particularly important role in organic chemistry; this group is called a carbonyl.

Basic Nomenclature of Organic Compounds

All organic compounds are made up of at least carbon and hydrogen. The most basic type of organic compound is one made up exclusively of sp3 carbons covalently bonded to other carbons and hydrogens through sigma bonds only. The generic name for this family of compounds is alkanes. Alkanes are part of a more general category of compounds known as hydrocarbons. Some hydrocarbons such as alkenes and alkynes contain sp2 or sp-hybridized carbon atoms. Alkanes are of great importance to the different classification systems and the naming of organic compounds because they consist of a carbon chain that forms the main structural unit of all organic substances. When an alkane carbon chain is modified in any way, even by the mere introduction of an sp2 carbon or a heteroatom (atoms other than carbon and hydrogen), is said to be functionalized. In other words, a functional group has been introduced and a new class of organic substances has been created. A functional group is a specific arrangement of certain atoms in an organic molecule that becomes the center of reactivity. That is, it is the portion of the structure that controls the reactivity of the entire molecule and much of its physical properties. An entire classification system of functional groups is based on atom hybridization.

Nomenclature of Alkenes

The IUPAC rules for naming alkenes are similar in many respects to those for naming alkanes.

1. Determine the root word by selecting the longest chain that contains the double bond and changes the

ending of the name of the alkane of identical length from ane to ene.

2. Number the chain so to include both carbon atoms of the double bond, and begin numbering at the end of the chain nearer the double bond. Designate the location of the double bond by using the number of the first atom of the double bond as prefix: Indicate the locations of this substituent groups by the numbers of the carbon atoms to which they are attached.

3. Two frequently encountered alkenyl groups are the vinyl group and the allyl group.

$CH_2 = CH-$ $CH_2 = CH\ CH_2-$

The vinyl group   The allyl group (are not included in IUPAC system)

The following examples illustrate how these names are employed

$CH_2 = CH - Br$ $CH_2 = CH - CH_2\ Cl$

4. The geometry of the double bond of a di-substituted alkene is designated with the prefixes, cis

and trans. If two identical groups are on the same side of the double bond, it is cis, if they are on opposite sides; it is trans.

Isomerism and its Organic Chemistry basics

The compound having the same molecular formula but differ in properties are known as isomers and the phenomenon is known as isomerism. There are two main types of isomerism.

Structural Isomerism – In this type of isomerism, compounds have same molecular formula but different structures. It can further be of following types:

(i) Chain Isomerism It arises when two or more compounds have the similar molecular formula but different carbon skeletons.

(ii) Position Isomerism- When two or more compounds have the same molecular formula but a different position of functional groups or substituents, they are called positional isomers and the phenomenon is called position isomerism.

(iii) Functional Isomerism – It arises when two or more compounds have the same molecular formula but a different functional group.

(iv) Metamerism – It arises due to different alkyl groups on either side of the same functional group in a molecule, e.g.,

(v) Tautomerism – It is a special type of functional isomerism which arises in carbonyl compounds containing α – H atom.

Stereoisomerism– The compounds having the same molecular formula but a different spatial arrangement of atoms or groups are called stereoisomers and the phenomenon is called stereoisomerism. Stereoisomerism is of three types: optical isomerism, geometrical isomerism, and conformations.

(i) Optical Isomerism – Compounds having similar physical and chemical properties but. differ only in behavior towards plane polarised light is called enantiomer & or optical isomers and the phenomenon is known as optical isomerism.

(ii) Geometrical Isomerism – The isomers having the same molecular formula but a different spatial arrangement of atoms about the double bond is known as geometrical isomers and this phenomenon is called geometrical isomerism.

(iii) Conformational Isomerism – In conformational isomerism because of the free rotation of carbon-carbon single bond, a different arrangement of atoms in space is obtained. These arrangements are called conformers.

Alcohols for biodiesel production

Alcohol is one of the most important raw materials for the production of biodiesel. Alcohols are primary and secondary monohydric aliphatic alcohols comprising 1–8 carbon atoms. A number of alcohols have been explored for biodiesel production, the most widely used acyl acceptors are methanol and to a slight extent, ethanol. Other alcohols utilized in producing biodiesel are the short-chain alcohols such as propanol, butanol, isopropanol, tert-butanol, branched alcohols and octanol, however these alcohols are costly.

Methanol and ethanol are the most often used alcohols in biodiesel production. Methanol is particularly preferred because of its physical and chemical advantages. Beside its reaction with triglycerides is quick and it can be easily dissolved in NaOH. Demirbas remarked that methanol, also known as "wood alcohol", is usually simpler to find compared to ethanol. Additionally triglycerides can react with varieties of alcohols. But the short-chain alcohols provide better conversions under the same reaction time.

CHAPTER 6

MAKING BIODIESEL

How to Build a Single Tank Biodiesel Processor

Now, almost everyone has heard of biodiesel, but few know that this environmentally friendly diesel fuel replacement can be made in a corner of a garage, using waste vegetable oil as the main ingredient. One only needs three basic ingredients to make biodiesel:

methanol (or ethanol, if you're a corn lover), oil (new or used), and lye (caustic soda). This book will tell you how to build a simple, straightforward single-tank biodiesel processor out of materials you may be able to round up for free.

Steps 1

Gather together the few common items listed under "Things You'll Need."

Step 2

Assemble the drum.

Cut a large opening (about half the top) in the top of the steel drum. You may also be able to find a drum with the top already removed.

Drill a 1.5-inch hole in the bottom of the drum. You will attach a pipe to it shortly.

Weld the 1.5-diameter pipe in the hole at the bottom of the drum.

Attach the 1.5-inch brass ball valve to the pipe. This is the drain valve. Brass is not necessary, but it offers the greatest durability.

Drill a hole in the side of the drum at the bottom, same size as the heater element. You can find heater elements in old hot water heaters or purchase them at a hardware store.

Fit the heater element properly, making sure it is not touching the side of the drum.

Wire up the heater element.

Step 3

Assemble the chemical mixer.

Attach one pulley to the rolled steel rod.

Attach the other pulley to the spindle of the electric motor.

Weld the propeller to the other end of the rolled steel rod. You can use cheap shelf brackets as propellers.

Attach the rod, pulley and propeller assembly to one side of the hinge. This will allow you to be able to move the propeller/mixer when necessary.

Weld a piece of angle iron across the top of the drum. This angle iron (or similar) offers support for the mixing apparatus.

Weld the unattached side of the hinge to the angle iron so the propeller and rod assembly sits in the middle of the drum. The hinge should swing the propeller and rod back and forth.

Mount the electric motor on the side of the drum. You should be able to mount the motor directly to the drum without it tipping the drum (unless your motor is from an old Cadillac or something).

Fit the belt to the pulleys and tighten by wedging a block of wood into the hinge. A tight belt is necessary to mix the biodiesel.

Tips

If you don't know how to weld, you can fashion a container that uses threaded pipe or industrial adhesives.

This is just one version of a simple biodiesel processor. You can use other items that you may find cheap or free. The basic idea is that you need a large, durable container, a mixer, and a way to heat the oil while mixing. Use your imagination!

Consider placing your entire processor on a stand with wheels. This will make your single stage processor portable and adaptable to most any space!

Think of your single tank processor as a large blender. One can easily make a blender batch of biodiesel, and the processor described above is simply a large rendition of a blender.

If you don't have access to a small motor, there are other options. One can use a drill with a propeller in the chuck to mix the biodiesel. There have even been models of single stage biodiesel processors that mount a bicycle on top of the drum, effectively stirring the biodiesel by mounting the stationary bike and peddling away!

It is essential to have a leak free container to ensure that the ingredients' proportions are correct and to keep messy vegetable oil from creating a slick out of your garage.

Warnings

Both ethanol and methanol are highly flammable liquids, and unfavorable conditions may result in flash combustion.

Place your processor in a well-ventilated area. The mixed chemicals will result in fumes that can be hazardous to your health. If the processor is going in a garage or outbuilding, make sure to place the processor near a door or a window that will open.

Sodium hydroxide can cause blindness in seconds. It hydrolyzes protein quickly, leading to severe eye damage. Cataracts, glaucoma, adhesion of the eyelid to the cornea, blindness, and loss of the eye may occur after eye exposure.

Research how to correctly and safely process biodiesel before making your first batch.

Lye (Sodium Hydroxide) is a strong base (alkaline) and will cause serious burns to the skin in a very short time and can cause blindness in SECONDS if splashed in the eyes. Chemistry-style goggles should be used whenever using Lye.

Make sure that your heating element is never in direct contact with flammable materials. Use the heating element to preheat oil prior to the addition of methanol or ethanol.

Flush exposed or irritated eyes with plain water or saline for at least 30 minutes. Remove contact lenses if easily removable without additional trauma to the eye, otherwise sodium hydroxide trapped beneath the lens will continue to damage the eye. If pain or injury is evident, continue irrigation while transferring the victim to the hospital.

Making biodiesel is addictive. Creating your own single stage processor may make you an oil baron.

If skin is exposed to Lye, it will first begin to itch, then a burning sensation will set in. If exposed, flush with cold water for several minutes.

Lye is a caustic ingredient, and contact with skin--especially wet skin—-will result in severe chemical burns.

Things You'll Need

55 gallon drum.

1/2 or 3/4 Hp (small) electric motor. An electric drill will also work.

Two pulleys which produce 250 rpm and a max of 750 rpm at mixer blade. You will not need the pulleys if you use a drill to power the propeller.

A belt for the above.

12 inch (30.5 cm) rolled steel rod. The rod can also be shorter or longer—it just needs sufficient propellers to mix the contents of the entire drum.

Two steel shelf brackets (for the blade).

1 1/2 inch (38mm) brass ball valve.

A hinge and a spring to act as a belt tension device.

2000-watt electric water heater element. Lesser or greater wattages will simply increase or decrease the time to pre-heat the oil.

A water heater thermostat.

1 1/2 diameter piece of steel pipe 3–5 inches (7.6–12.7 cm) long with male threads on one end.

The Raw materials For Biodiesel

Catalysts for Biodiesel

A key ingredient in the homebrew biodiesel recipe is catalyst. Sodium Hydroxide and Potassium Hydroxide are the most commonly used. The glycerin and soap that settles out from a reaction using Sodium Hydroxide will be solid, while the soap and glycerin that settles out from a reaction using Potassium Hydroxide will be liquid.

Methanol for Biodiesel

Methanol or wood grain alcohol is the simplest form of alcohol. It is both poisonous and flammable. It is used in racing, making it available at both chemical supply houses and fuel distributors. It burns without a visible flame in bright sunlight, and it is the most expensive ingredient for most homeb rewers.

Oil for Making Biodiesel

There is a huge variation in fats and oils used to make biodiesel. This article goes into the properties of the different oils to help you know which are best for making biodiesel. For instance, the saturated fat content of oil is a strong indicator of its cold weather performance.

How to Collect Waste Vegetable Oil

Cheap and free oil is what makes homebrew economical. Homebrewers have a big advantage over commercial producers because they can pick up oil on their way to and from work. Commercial producers either have to buy their oil or pay to have it delivered or have to hire a driver to collect oil directly from restaurants.

Drying WVO for Biodiesel

The most common cause of problems when making biodiesel is water in the feedstock oil.

Water Washing Biodiesel

Washing Biodiesel with water is the oldest and most common method of cleaning biodiesel. About 3% of raw, unwashed biodiesel is methanol. Methanol is a solvent, it captures soap and other impurities and holds them dissolved in the biodiesel. Water soaks up that methanol, releasing impurities to be washed away with water.

Keeping the methanol liquid and diluted in water makes water washing the safest way to clean biodiesel. Most dry wash methods require we evaporate or distill the methanol into a flammable and toxic gas as part of the purification process.

Water washing is the most flexible way to purify biodiesel. Under the right conditions, you can power wash in just a few hours with extremely aggressive wash methods. Alternatively, you can take up to a week using less aggressive methods. Some methods conserve water, letting you wash with 1 part water for every 5 parts biodiesel, while most use a 1:1 water to biodiesel ratio. Water washing can be automated, and you can mix and match different wash methods to fit your personal needs.

Water Wash Methods

It takes several washes to clean your biodiesel. Most people start with a less aggressive wash method, moving on to more aggressive methods with later wash cycles. I like to start with an overnight Static Wash, then move on to bubble washing. With each wash, I put one part water for 5 parts biodiesel. If I have a 40-gallon batch, I replace the water with 8 gallons of fresh water for each wash. The first two bubble washes usually last about 8 hours each, with the last water wash usually lasting 48 hours. I use a lamp and appliance timer to control the bubble wash, so I only need to attend to it for a few minutes a day.

Static Washing

Static or gravity washing is the least aggressive and least likely to generate an emulsion. It is simply placing water and biodiesel in the same tank without any mixing. Impurities migrate from the biodiesel to the water through the boundary layer over time. This process takes anywhere from 4 hours to 24 hours to saturate the water with contaminants. Most homebrewers let the static wash continue overnight before draining and starting a different technique. It is particularly effective as the first wash on very soapy biodiesel made from high FFA oils.

Appearance of the Wash Water

It takes some experience to read this test properly. If you wash with soft water, the water on the bottom is always soapy. If you have hard water, it may be clear when there is still soap in the biodiesel. Different oils can also give different readings. Even though it is subjective, it can still be a useful quick test, once you have sufficient experience with it.

Wash Water Treatment and Disposal

Many of us have been disposing of our wash water by either pouring it out on the ground or down the sewer. The "Man" is likely to have a cow if he finds out you are doing that. If any gets into a stream or goes down a storm drain, then you are in violation of the clean water act. If you pour it down the drain on a city sewer, you are probably overloading the system and causing the wastewater treatment plant to get fined by the federal government for discharging untreated sewage.

The wash water contains both soap and biodiesel. The soaps are heavy in carbons, which demand large amounts of oxygen to biodegrade. The term to describe this is Biochemical Oxygen Demand (BOD). Typically, the wash water from the first wash usually kills your lawn. If you use NaOH, the second and third washes can also kill the lawn. One way to reduce the BOD is to let the wash water settle in a drum for a week or so. Over time, some of the biodiesel bound up in the soaps rises to the top of the drum.

Biodiesel Safety Tips

Biodiesel Safety Tips Over the years there have been a number of biodiesel processing related accidents and fires, even some serious injuries and a couple of deaths. Making biodiesel is inherently dangerous, you are working with toxic flammable liquids and strong caustics to make fuel. While we cannot eliminate all the dangers, here are 10 basic safety tips that can help reduce your exposure to accidents and fires.

Tip 1: Stay awake, sober and alert.

Your brno sleeping at workain is your best tool for preventing accidents. If it's asleep or impaired, or just not being used, then you're an accident waiting to happen. If you were to fall asleep at the wheel and kill yourself in a car crash, everyone would consider it suicide. Making biodiesel is no different. Both activities can be lethal if you attempt them while you are not in full control of your body. The same principal applies to making biodiesel while drunk, it's suicidal, don't do it.

Tip 2: Do not process in your home.

biodiesel fire If you don't burn down your home, you can poison your family with methanol vapors. It's next to impossible to make biodiesel without releasing methanol vapors. If you process in your basement or attached garage, those vapors will work their way into your bedroom or your children's bedroom. The effects of methanol poisoning are cumulative. Repeated exposure at low levels can cause chronic symptoms to appear. It's even more dangerous to children as their small bodies can tolerate less poison than adults.

This is fuel we are making. It may be "safer" than diesel due to it's higher flash point, but it's still fuel. It will make any fire burn hotter and faster. In fact, once your oil and biodiesel start to burn, you only have a few minutes before it totally destroys your home. In most of the fires involving processors in the home, the house was a total loss before the fire department could even arrive. If you do decide to keep biodiesel or oil in your home, then follow the local codes for Home Heating Oil (HHO) storage.

Tip 3: Keep your work area clean and free of oily rags.

fire destroys biodiesel shed Spontaneous Combustion is one of the leading causes of fires in home brewing operations according to fire investigators. Oily rags will burst into flames without an apparent ignition source if the conditions are right. Unfortunately the right conditions are fairly easy to replicate. Just pile up some oily rags. The oil will chemically react with the air in a process called oxidation, giving off heat. The rags act like insulation holding in the heat, letting it builds up to auto ignition temperatures.

Tip 4: Don't stockpile glycerin.

Many home brewers don't have a plan to dispose of their glycerin. Often they will put it in the plastic containers they collect oil in. Then just stack it up. Those containers become brittle and develop leaks when exposed to sunlight. The methanol in the glycerin will attack the thin hope containers and cause them to leak.

Biodiesel and vegetable oil in small quantities can be handled safely, but hundreds of gallons not only pose not only a fire hazard, but also an environmental hazard.

Tip 5: Use timers on heating elements.

biodiesel equipment timer You never know when you may have to leave your processing area. Emergencies happen. If heating elements are left on while you're away in town, you run the risk of overheating your oil. In fact over heating oil on a stove is the most common cause of house fires.

The smoke point of oil is the point at which it starts to smoke and becomes a fire hazard. That's when it starts to break down into low flash point compounds. Over time as oil is heated for cooking it's smoke point will drop. Restaurants on a tight budget will use cooking oil until it starts to smoke in the fryer. That puts smoke points at under 350F for typical WVO used in homebrew. In fact anything that causes titrations to go up also causes smoke point to go down. High titration oils can have smoke points as low as 200F. We can easily reach these smoke point temperatures in insulated drums with the small heating elements we use in processing.

Tip 6: Use Secondary Containment.

fire consumes biodiesel barnIf you have a major oil leak, secondary containment can reduce both environmental and fire hazards. It is used by commercial brewers as cheap insurance against big spills reaching rivers and streams. Cheap that is when compared to the cleanup costs and the fines slapped on polluters.

For home brewers it keeps spills contained until they can be cleaned up. When an oil spill soaks into the dirt below your shed, it will ⬜uickly go rancid. The only way to get rid of that awful smell is to tear down the shed so you can dig up and haul off the contaminated soil.

Tip 8: Don't use a drill and paint stirrer for mixing

sparky drills make methanol vapors explode Back in the bad old days we used drills and paint stirrers to mix biodiesel. It was a cheap easy way to mix both biodiesel and methoxide. The problem is that it's excessively dangerous. When methanol vapor concentrations around the drill reach 6%, the sparks from the drill will ignite the methanol. It will form a fireball centered on the drill. Since your hands are holding onto the drill they receive 3rd degree burns as do your forearms.

Tip No 9: Don't make test batches in a blender

making biodiesel in a blender can be dangerous You see it all over the internet, in videos and websites, people making small batches of biodiesel in a food processing blender. The practice goes back to the beginning of homebrew biodiesel and it's time for it to stop. Food processors and blenders were designed to process food.

Tip No: Don't make biodiesel in a closed head drum

Closed head or tight head drums are not designed to be pressurized. They will come apart at the seams violently if over pressurized. One way you can over pressurize a drum is to ignite the methanol fumes by uncovering an electric heating element while it is on.

STEPS TO MAKING BIODIESEL

1. The base amount of lye catalyst needed

For clean, un-used vegetable oil, there is a base amount of catalyst which will need to be dissolved into the methanol to make a complete reaction. For NaOH, the commonly accepted amount is 5 grams of NaOH per 1 liter of vegetable oil to be converted. Since KOH is less dense, it requires 1.4025 times as much, which is 7 grams of KOH per 1 liter of vegetable oil to be converted.

Purity must also be taken into account. Usually with NaOH, purity can be neglected, since it is often nearly 100%. However, it is important to factor in purity with KOH since it is usually about 90%. To factor in purity in your calculations, divide the base catalyst amount by the % purity. We usually carry NAOH with a purity of 99.1% or so, which makes the calculation 5g/0.99 = 5.05g of NaOH per liter of oil. For KOH, it's usually about 90%, which calculates as 7g/0.90 = 7.77g of KOH per liter of oil. Check the current certificate of analysis for the catalyst you are using to calculate the most accurate amounts possible. Usually, going with the 99% or 90% is sufficient since the differences in hundredths or even tenths of a decimal can be lost in the weighing process, especially when doing larger batches.

When converting WVO (waste vegetable oil) to biodiesel, more catalyst is needed than for clean oil. The recipe will require the base catalyst plus an additional amount which must be determined by a titration each time a different batch of biodiesel is made.

2. Titrating waste vegetable oil (WVO)

Waste vegetable oil (WVO) containers Free fatty acids (FFA) which cause vegetable oil to be viscous and even gel at room temperature. The more FFA an oil has, the more lye catalyst you will need to replace the FFA with methyl esters.

Since FFA content increases with continued use in a fryer at a restaurant, you must perform a titration to find out the FFA content. To do this, you will need the following:

- 99+% Isopropyl Alcohol

- 1% Phenolphthalein Solution in alcohol

- Accurate Syringes

- Beakers (50 ml is ideal) or small containers to mix the solution in

- 1 liter (or larger) bottle

- distilled water

- Lye Catalyst (sodium hydroxide or potassium hydroxide)

- Pocket Scale accurate to at least 0.1 grams, preferably 0.01 grams.

The process:

Create the 0.1% by weight Lye Catalyst Solution

First, you will need to make a 0.1% by weight catalyst solution in distilled water. To do this, measure out exactly 1 gram of sodium hydroxide or potassium hydroxide, depending on which catalyst you will be using in your process. Dissolve the 1 gram of catalyst into 1 liter of distilled water. For most accuracy, weigh your water rather than measuring it. 1 liter of water weighs 1 kg. it should actually be 1g of catalyst in 999ml/999g of water, but most are unable to be exact on such a measurement for such a small correction in error since scales that weigh up to a kg usually do not weigh accurate to the tenth of a gram.

Should your measuring devices have an accuracy concern, proportionally increase the amount of your solution. For example, add 3 grams of catalyst into 3 liters of distilled water. Making larger measurements will leave less room for error, but may not be worth it if you don't plan to use that much solution.

Once you have a 0.1% lye catalyst solution prepared, you will be able to perform your titration. Catalyst solutions are only good for about a month before they begin to weaken. Make a fresh solution every month.

Perform a blank titration

It is important to make sure that your isopropyl alcohol (IPA) is fresh. Since IPA does not have a ph rating, you cannot simply test it with a ph meter. However, you can perform a blank titration. To do so, measure out 10 ml of IPA and place it into your beaker. Add a couple drops of phenolphthalein. Next, use a 3ml or 5 ml syringe to measure out your lye catalyst solution. Put a drop or two at a time into the isopropyl alcohol. It should turn magenta within a few drops, usually on the first one. If it takes more than a few drops to turn magenta, your IPA is bad and you should replace it. You still can use it however, but you must make sure to perform the blank titration up until it begins to turn a light magenta before starting the titration process. If your IPA is good, you will not need to perform this step again unless the IPA has been stored for a long period of time or you suspect contamination.

Performing the Titration

Use a 10 ml syringe to measure out 10 ml of IPA. It is preferred to use at least 99% industrial grade or better IPA, since the water content in dilute solutions will affect results. Put the 10 ml of IPA into a small beaker, 50 ml beakers work the best. Add a couple of drops of the 1% phenolphthalein in alcohol solution.

Measure out exactly 1 ml of your WVO sample using a 1 ml syringe which is accurate to at least a 10th of a ml, preferably accurate to a 100th of a ml. Dispense the 1 ml of WVO into the beaker with the IPA. Swirl it around a little by moving the beaker to ensure it dissolves completely into the IPA. It could be a white cloudy mixture or almost clear.

There are two methods for measuring the titration process. Using a pocket scaleaccurate to at least a tenth of a gram is the best way. Place the beaker onto the scale and tare the weight. After the titration, you can place the beaker back onto the scale to see how many grams of catalyst solution you have added. 1 gram of water = 1 ml of water. The other method is to use syringes. usually, a 3ml syringe works fine for NaOH and a 5 ml for KOH, depending on titration levels.

If you have a magnetic stirrer, this is a great time to use it. Use the stirrer to gently stir the mixture in the beaker, or if one is not available, gently stir the mixture with a stirring rod or by swirling it with movement of the beaker. next, slowly dispense the catalyst solution into the beaker while mixing it. It will usually turn a cloudy white color and then begin to show signs of pink. when you start seeing the pink, add the catalyst solution drop by drop until it remains that pink/magenta color for 30 seconds. Once you have achieved this result, you are ready to measure your results and calculate the titration amounts.

Calculating lye catalyst needed from titration results.

Count how many grams (or ml for syringes) of catalyst solution it took to complete your titration. Each gram/ml of solution required represents 1 gram of lye catalyst needed in addition to the base catalyst to process your WVO. base catalyst + ml needed to titrate = total grams needed to process the oil.

As an example:

Let's say it takes 2.3 ml of lye catalyst solution to complete your titration. If you are using sodium hydroxide, the base catalyst needed is 5 grams. So that's 5g + 2.3 g = 7.3g of sodium hydroxide per liter of oil. Divide the base catalyst by 0.99 and then add the 2.3g to get 7.35g for purity corrections. As you can see, it doesn't really matter much with NaOH but it's a good practice. Be sure to only factor purity into the base catalyst. If you're using the same catalyst in your catalyst solution, purity corrections have already been made within the mixture itself.

For potassium hydroxide, the base catalyst is 7 g. So that's 7g + 2.3g = 9.3g per liter of oil. Don't forget that potassium hydroxide is not normally anywhere near 100% pure, so you will need to factor in the purity by dividing the base catalyst by the purity. Assuming 91% purity, it would be: 7g/0.91 + 2.3g = 9.99g of KOH needed per liter of oil.

3. Methanol

Methanol, like the catalysts, also absorbs moisture from the air. Be sure that when working with methanol, to work quickly and seal the containers in between uses as soon as possible. The less moisture you have in your process, the less soaps you'll end up with in the finished product.

The amount of methanol required for the process is about 20% of the volume of the oil to be processed. Some brewers use 21% or even 22% to be sure there is enough. To be clear of how much methanol to use, if you were to process 100 liters of vegetable oil, the process would require 20 liters of methanol for 20%.

Once the methanol has been measured out, the catalyst now needs to be mixed into a solution with the methanol. For clean oil, use the base catalyst amount and for used vegetable oil, use the base catalyst + titration amount.

The reaction between methanol and the catalyst is exothermic. It will release a large amount of heat during the reaction. In the case of plastic tanks used in processors, the amount of heat released is usually not enough to make the methanol boil off, but it is good practice to add half first, let some heat bleed off and then add the rest if a lot of catalyst is needed.

NaOH releases more heat in the dissolving into methanol. KOH tends to be a lot cooler when dissolving into the methanol and can usually be mixed immediately and be ready to go in minutes.

For small amounts of mixtures of methanol and catalyst, it can be mixed together quite easily by putting the methanol and catalyst into a sealed container and then shaking and swirling it around. Once all of the catalyst is dissolved, it's ready. For larger applications, it's best to attach a propeller to a stainless steel shaft which can connect to a power drill. The drill can be run for only about a minute and there should be enough agitation to fully dissolve the catalyst into the methanol. Since methanol is flammable and drills can spark, this shaft should be attached through the cover to the tank as a permanent attachment. We have ours installed with ball bearings inside of the cover for an easy spin.

Another method would be to use a pump, drawing the methanol from the bottom of the tank to the top of the container, with a screen blocking the catalyst from flowing through the bottom until dissolved. Some processors use the mixing pump for the biodiesel reaction to perform the methanol mixing task by the switching of a few ball valves. it works, but it will get some biodiesel/oil and maybe glycerin into your methanol mixing tank.

4. The Process of a test Batch

For first-time users, it is best to do a few test batches first to be sure that measurements are being made correctly and to see how the process works. If a new brewer jumps right to the processor, lack of experience can lead to a processor full of badly reacted biodiesel and a lot of headaches.

On a small-scale, accuracy in weights and volumes of the vegetable oil and catalyst is key, so if a user can master using the right amounts for a small batch, larger batches can easily be made since there is more room for error in a large batch. Most test batches process 1 liter of vegetable oil, which requires a standard 200 ml of methanol with 5g of naoh or 7g KOH base catalyst (purity not factored in) + titration amount for used oil. First-time users should start with fresh oil so the reaction alone can be familiarized with, and then move onto waste vegetable oil to practice processing successfully with a titration involved. If you change your type of process at any time, such as changing from naoh to KOH or even changing to a different oil from a different restaurant, it is best to perform a test batch before a larger batch.

1 Liter Test Batches

For a small batch of biodiesel (test batch) usually a cheap blender is an easy way to perform the mixing process. Heat the 1 liter of oil up to about 140°F (60°C) using an old pot on a stove. Don't let it get any hotter or you will end up boiling your methanol when you add it.

Prepare your methanol/catalyst mix by mixing the methanol and required catalyst in a compatible container. Be careful not to use just any plastic like a water bottle as the mixture is corrosive and is likely to eat the bottle away. Stick to tough plastics such as HDPE (High density polyethylene). Keep the container closed while mixing, it can be shook vigorously and the cap partially unscrewed in between shakes to release the pressure from the gases created in the reaction. CAUTION: Do not breathe in these gases. They are toxic! Once the catalyst has been fully dissolved into the methanol, you are ready for the reaction.

A magnetic stirrer would be ideal for the mixing process, however, most people starting out will probably want to use an old blender. It's best if the blender is made of glass and not plastic. Plastic ones are usually cheap and may crack under the chemical mixes and heat.

Pour the oil carefully into the blender, cover with the cover and then start the mixing on a low speed while holding down on the cover. Once all the initial turbulence has settled down in the mixing, remove the cover and then slowly pour the methanol/catalyst mixture into the oil. Cover the blender again and turn on to a higher speed.

Agitation in a blender is plenty enough for making biodiesel. You should only need to run it for about 15-20 minutes. Once the mixing is completed, stop the blender and pour the mixture into another container, preferably clear HDPE for watching the glycerin by-product drop out. Don't leave it in the blender. The regular rubber gaskets will deteriorate over time when exposed to biodiesel. You can expect the blender to only last for a few test batches and to never be used again for anything else.

Whichever mixing method you use, after about an hour, most of the glycerin will drop out of your product and settle to the bottom. it will be a darker color and appear much more viscous than the clearer biodiesel on the top. You should wait at least 12 hours before washing it, preferably 24 hours. The longer you wait, the more glycerin will drop out and the easier the wash will be. Some users will wait 2-3 weeks to let all impurities drop out, and then the water wash will be perfect and clean.

Washing a test batch

Washing is the most difficult part of the biodiesel process. It is extremely important to wash out all of the impurities left over in biodiesel that do not drop out with the glycerin. This includes soaps, salts and other such non-filter friendly contaminants.

Water washing is the simplest method for washing biodiesel but can be the most difficult to perform successfully especially with heavily contaminated batches. There are some dry washing methods available which are much more convenient and effective than water washing. The best dry wash method is the use of ion-exchange resin such as Dudalite in a dry wash tower. Another available dry wash media is Magnesol. for washing a test batch, it is best to stick to the simple water wash method and then move onto the dry wash media on larger batches.

To water wash your test batch, drain off the glycerin from the bottom. Most likely, you don't have a drain on the small container you put the biodiesel into. Even if there is a drain, it will not be convenient like in a larger batch. You will be likely to end up draining out half of the biodiesel before you can get all of the glycerin out of the bottom. What I do is pour the top part (the biodiesel) into another container and then pour the bottom part (mostly glycerin but still some biodiesel) into one of our tri-pour beakers. Once resettled in the beaker, almost all of the biodiesel can be poured into the container, leaving all of the glycerin set aside in the beaker.

Next, once you have just biodiesel and NO glycerine, add 1/3 as much water as biodiesel in a slow manner so as not to agitate the biodiesel too much. the water will sink to the bottom quite readily if it's not agitated. Now, turn the bottom upside down and let the water settle down to the top, which is now the bottom upside down. And continue to do this over and over until the water becomes very murky. Be careful not to agitate the biodiesel too much, or you will have to wait a long time for it to separate, possibly up to a week or longer.

After the water is dirty, it has captured a lot of the excess methanol, soaps and other contaminates in the biodiesel. Let it settle for a good amount of time so that you have all biodiesel on the top. if you are gentle enough, it should only need about 15-30 minutes to completely settle out. Proceed to drain out the biodiesel into another container as you did before when removing the glycerin, and then add 1/3 water again and wash again. Each time you wash, the water should start to become less and less murky. As the biodiesel gets cleaner, you will also notice that you can be a little more vigorous in the washing. Be careful not to agitate it too much so you don't have to wait too long for it to settle. The goal is to get to a point where the water no longer gets cloudy at all and remains clear. This happens when the biodiesel has been washed completely. usually this is achieved after 3-4 washes. The wash water should be at a neutral ph of 7 when the biodiesel is completely washed.

Drying a test batch

Although properly washed, biodiesel will not appear fully clear until it has been dried of all of the water. this can be achieved in a few different ways. The easiest but longest way is to set the container out into the sun so the water can slowly evaporate out. It can take up to 3 weeks for this process to be complete. To speed up drying for small test batches, you can pour the biodiesel into a pot and gently heat it up to 212 degrees. Above its boiling temperature, water cannot exist in liquid form. So once your biodiesel has passed 212°F and the bubbling has stopped, there is no need for additional heat as all of the water will have boiled off. Some people like to heat to something like 250°F to be sure, but this is a bad practice and a waste of energy since it is impossible for water to exist beyond 212°F at sea level.

When the final biodiesel product has been fully dried, it will be a nice clear or almost translucent color. This is your finished product, ready to be filtered and filled to your fuel tank for use.

5. Processing Larger Batches of Biodiesel

Once you have mastered 1 liter test batches of biodiesel, it will be time to start thinking more about the processor than the process. Building a processor is ☐uite simple, and you don't need all of the fancy looks of expensive processors to get the job done.

Building your own processor

It is most common for biodiesel makers to use conical bottom shaped tanks for the purpose of easy draining and separation of by-products from the biodiesel. The most common set-up is to attach a valve to the bottom drain which leads to some plumbing going to a circulation pump and also to another drain which can be used to actually drain off the bottom. I find it best to attach that drain to a hose so that by-products and biodiesel can be drained to any container. Some people even attach a 2nd pump for pumping the finished biodiesel out of the tank.

The 1st pump should be plumbed in so that it can take biodiesel and anything mixed in with it such as biodiesel and pump it up to the top of the tank. This is the best mixing method of the processor. This pump should be powerful and ideally should be able to circulate 1/3 of the tank within a minute's time. the slower the pump, the less agitation you have and the longer the processor will need to run.

Circulate the batch from the bottom and back to the top in a continuous loop. It is often a common practice to install temperature gauges or thermostats into this part of the plumbing as well as a water heating element for temperature control. We install plate heat exchangers into this part of the line so we can use the hot water coming from our solar water heater. We are all about being green and efficient, and if you want to truly be green, i highly suggest adopting solar water heaters into your process. Not only will it save you a lot of energy on biodiesel making and drying, but it will also take away from your energy bill when it comes to using hot water elsewhere in your home.

A 2nd smaller tank is often plumbed in line with the 1st tank for mixing the methanol-catalyst solution and then slowly adding it to the process as vegetable oil is pumped from the bigger tank's bottom and back to the top. the best way to do this is to run the bottom of the methanol-catalyst solution tank through a small pump and down to a tee just before the circulation pump. Use a one way check valve right at the connection point to ensure that vegetable oil doesn't find its way up into your methanol-catalyst solution tank from the pressure created by a lot of fluid in the larger tank and gravity overpowering the circulation pump's ability to suck in both the vegetable oil and methanol-catalyst solution at the same time. the purpose of the methanol-catalyst solution pump is to help push the methanol-catalyst solution through that check valve to completely empty the methanol-catalyst solution tank. it should also be as close to the check valve as possible and preferably with a little bit of head pressure to help push it. A ball valve should also be installed on the drain of the methanol-catalyst solution tank so that the methanol can mix evenly with the catalyst prior to dispensing it.

Mixing the methanol-catalyst solution can be tricky. Since you're dealing with larger amounts of methanol, you cannot simply shake it up like in a test batch and sometimes there's too much catalyst required to just let it sit overnight and dissolve. The best solution to this, is to find a propeller like what you find on a boat motor, run it to a shaft which comes up and through the cover in the tank and is sealed off with some washers so that it's free to rotate, and then use a power drill to attach to the shaft and spin the propeller at rapid speeds. Amazingly, it will only take less than a minute for KOH to dissolve into methanol with this set-up, if even that long. You should also attach a fine screen on the top of the drain of the tank to prevent the catalyst from falling into the plumbing.

This usually isn't much of a problem though if you have a ball valve, but is still recommended. be sure the propeller does not touch the screen.

Sizing your processor

the methanol-catalyst solution mixing tank needs to be at least 20% the size of the batches you are planning to do, and the mixing tank needs to be at least 120% the size of the batches you are doing to account for the vegetable oil and methanol added. Most standard processors have a 60 gallon mixing tank with a 15 gallon tank for mixing methanol-catalyst solution Other sizes could be an 85 gallon tank with maybe a 15 gallon methanol-catalyst solution tank or a 110 gallon tank with a 30 gallon methanol-catalyst solution tank. for the 60 gallon the largest possible batch would be 45 gallons of vegetable oil and for the 85 gallon it's 65 gallons and 110 gallon would be 85 gallons. since the tanks' max capacity is based on the top most height of the tank however, it's usually best to process about 5 or 10 gallons less than these max capacities in order to avoid spillage from overflow during pumping or when there is a bad measurement.

Materials for your processor

HDPE and other resistant to biodiesel/caustic/methanol materials should be used when building a processor. Stainless steel is the absolute best metal to use and regular steel, galvanized steel and especially copper and aluminum should be avoided since they will corrode quickly and affect the quality of the process. PVC pipe can be used, but it's best to stick to threaded piping since glued pieces can spring leaks after awhile and cause more headache than it's worth.

Regular rubber hoses cannot be used. Biodiesel quickly degrades the quality of rubber and rubber hoses would need to be replaced often. Clear vinyl hose tends to be the hose of choice since it holds up against biodiesel for quite some time and you can see the fluid inside of it. This is especially great for the drain hoses. Fluoroelastomer hose will last the longest when exposed to biodiesel and processing biodiesel, however, it is not so good when up against the pure methanol-catalyst solution since methanol degrades grade a Fluoroelastomer hose. it is a very good idea to use vinyl hose when connecting from the methanol-catalyst solution tank to the plumbing of the larger mixing tank so you know when it has been fully emptied.

Moisture Removal

It is extremely important to make sure there is no water in your vegetable oil before processing. Depending on how often you need to process biodiesel, you may need a few settling tanks for removal of water. Settling is the easiest way to remove water. If vegetable oil is left in a cone tank for 2-3 weeks and left untouched, all of the water can be drained out the bottom readily. This process can be sped up by installing a heating element at the bottom of the tank and gently heating the veggie oil up to and maintaining it at about 100°F to 110°F for 4-6 hours. The heat decreases the density of the vegetable oil greatly while the water's density remains about the same, settling the water down faster than normal.

After the initial 4-6 hours of maintaining heat, the vegetable oil should be completely separated after 12 hours, preferably 24 hours. use the crackle test to ensure no water is present once you have removed the water from the bottom.

Processing with your own Processor

No matter how powerful your pump is, you will not achieve as much agitation as in a blender or magnetic stirrer. You could use a blender like set-up instead of a pumping system, but blades are likely to degrade when left in the biodiesel for a long period of time and it's not so easy to set such a system up. There is also a better chance of spillage when this method is used, and you won't be able to access the top of the processor very easily when you need to.

The time it takes to successfully process a large batch of biodiesel will depend on the power of your pump. Generally, it takes about 1.5 to 2 hours of circulation to fully complete the reaction. Some users need to run it for as long as 4 hours and others can get away with shorter times. Heating of the biodiesel is also very important. The temperature should be maintained at about 150°F for the entire mixing process. if lower temperatures are used, it will require longer mixing times.

Once the process has been completed, you can stop the pump and turn off the heat. let the biodiesel settle overnight just like you would with a test batch. Settling is mostly done in 12 hours, but you should wait 24 hours to be sure. the longer you wait, the better.

6. 3/27 Methanol Test / 10% Biodiesel Test - Testing for a Complete Reaction

How to Know When the Biodiesel Reaction is Complete

Biodiesel dissolves quite readily into biodiesel while vegetable oils or fats do not. The 3/27 Methanol test, or 10% biodiesel/methanol test can be used to determine whether a reaction in your process is complete and you are able to turn off the process to settle out the glycerine.

Performing the 10% Biodiesel/Methanol Test

1. Take a small sample from your processing batch into a beaker and let it settle for a few minutes.

2. Draw exactly 27 ml of methanol using a syringe and place it into a clear, thin glass jar with a good lid which is able to hold at least 30ml of fluid but not too much more than 100ml.

3. Once the glycerin in the biodiesel sample has appeared to mostly settle out, draw exactly 3 ml from the top and add it to the 27ml of methanol in the jar to make a 30 ml solution.

4. Shake up the solution vigorously for a few seconds and then place it onto a level surface for observance. Within ten seconds, the solution should clear and you should be able to see right through it. Most likely if it is perfectly clear, the reaction is complete.

5. Inspect the solution for non-miscible parts of oil floating in the solution. If you can see any dots or traces of oil, typically floating on top of the solution or possibly lingering at the bottom, the reaction is not complete. Usually solutions with spots in them will also be cloudy, so when it is completely clear, it is

often a safe bet that the process is complete. Some complete batches may not be of the highest quality and will be a little hazey but still have no floating oil.

If your solution is clear and without any oil spots, your reaction is complete and you may shut off your processor and begin the settling process. If there are oil spots found, leaving your processor on for further mixing and test again. continue processing and testing until you reach a clear, unspotted solution. After performing this method a few times on a few batches, you will get a good idea of exactly how long each batch should take.

7. Washing Biodiesel

Clean fuel is essential for keeping an engine away from maintenance later on. Washing biodiesel, whether by dry wash resin or with water will remove soaps and other contaminates. This is essential to ensure your fuel filters will not get clogged and also to be sure your engine will not get beat up. If using water washing, it is best to remove the methanol first prior to doing the wash. If using dry wash resin, whether you remove the methanol first or not will depend on your soap content. Do a soap titration on your fuel prior to the dry wash and determine if utilizing a methanol wash is necessary or not.

Washing is easiest when all possible glycerin has been removed. Glycerin drops out fastest when excessive methanol has been removed from the biodiesel. This can be taken out by one of two ways, distillation (methanol recovery) or by simply evaporating it into the atmosphere. Distillation is covered later under methanol recovery. If you do not wish to recover the methanol for re-use, you can simply bubble the biodiesel with a fish tank compressor, while blowing a fan above the tank to help force the air with methanol fumes out of it. The methanol should be completely gone after about 24 hours. Note that this cannot be done in a closed area and the methanol fumes should be vented outside. Once the methanol is removed, excess glycerin should rapidly drop out in about 24-48 hours.

If you use a full drain tank, meaning, there is no bulk head, you will be able to fully drain your glycerin and you can potentially use your mixing tank as a wash tank when washing with water. Even though this may work, it is recommended to have a separate tank from water washing. If you mess up, you're stuck with waiting until your batch is fixed before you can do another, and you will need to wait for the tank to dry after using it before you can use it on a new batch. If you are dry washing, you can pull the biodiesel right from the tank, through your towers and into your clean storage tank.

Assuming you transfer the biodiesel to a wash tank, drain off the glycerin into a storage container and set it aside, then transfer the biodiesel to your wash tank. there are several ways to perform a wash, and depending on which one you use will determine how you set this tank up. do not try to just pump from bottom to top like you did with the mixing, that's too much agitation and it will make a mess.

Mist-Washing Method

Mist washing is done by installing misting nozzles at the top of the wash tank and slowly introducing water through them and into the biodiesel. The fine spray of water slowly travels down through the biodiesel, taking contaminates with it. After the tank fills up from the water, usually with about 1/3 worth of water, you can drain off the water and then continue spraying. the process will need to be completed until the water comes out clear, which means there are no contaminates left.

Mist washing uses more water than bubble washing because it is not circulated continuously through the biodiesel, however, it is a little more convenient since there is less waiting involved since the biodiesel doesn't need so much time to settle out except for in the final draining of the water.

What to do with your wash water

Wash water should not be dumped down the drain, especially if you have a septic tank installed. The methanol, soaps and other contaminates pulled out of the biodiesel can be hazardous to marine life and will kill the bacteria needed in a septic system. Check with local laws and codes about disposal of your wash water. Some larger operations tend to have a holding tank outside which can dry in the sun and once the water has all evaporated, the solids can be scooped out and disposed of into a landfill, according to local laws. if you do this, make sure your tank/wash water holding area is much wider than it is tall. you will need as much surface area contact with the sun as possible for fast evaporation. Water takes a lot of energy to heat, and even more to evaporate. If you live in a rainy area, this is probably not the best way to handle this.

We used to rid of our wash water through evaporation by using coolant from our bus engine while driving and circulating it through a radiator in a metal tank. Since most people are doing this process at home, this isn't really an option, however, it brings out other ideas as described below.

Another option to consider would be to again utilize a solar water heater.The best way to do this is to get a metal drum or some sort of temperature resistant container, drop an old radiator into it with the wash water, and circulate the hot water from your water tank or even fluids directly from the solar heater through the radiator. It will heat up the water throughout the day and evaporate it out much more ⬚uickly than just plain sunlight.

other users may find it convenient to boil the water out of a drum by burning a fire under the drum or using heat from a boiler or wood stove with a medium fluid similar to the set-up for the solar water heater. Most methods will work if it involves adding heat to the water for a good amount of time, but i only really want to promote the greener way of doing it. Other methods that expend energy may save time but are essentially wasteful. If you don't have the patience for getting out the water to save money on the wash water disposal, i would recommend dry washing instead.

Two-Stage Process (Acid/Base)

WVO with high FFA content is difficult to convert to biodiesel using the single stage base process. Many users will find that as titration levels get higher and higher, there will be more and more glycerin byproduct and less fuel yied. If the FFA content is too high, the conversion will not be successful, even with the best and most careful process. For this reason, the acid/base two-stage process was developed in order to convert higher titrating oils. This process is usually intended for titration levels of 12 or more (using

KOH) but is also very successful for lower titration numbers and even nearly clean oil. .

Here at Duda Diesel, we've found the two-stage process to be the best way to go for experienced users. It makes a higher quality biodiesel, and will generate the greatest yield. There also appears to be less soap generation, which is likely due to the fact that less base catalyst is required. Once a user is familiar with the two-stage process, the process can also save a lot of time on the mixing part of the process. The two stage process is sometimes referred to as the Fool Proof method to making biodiesel. While it makes the process easier, one who does not make proper measurements or overestimates the amount of acid required wastes chemicals and will end up with a lot of unnecessary salts and water as a byproduct of the process. However, overshooting with too much acid will not ruin a batch, unlike in the one stage process where too much base catalyst will make soap.

The two stage process starts with concentrated sulfuric acid. The sulfuric acid acts as a catalyst and allows for the FFA to undergo esterfication prior to the transesterfication process with the base method.

By converting all of the FFA using the sulfuric acid, theoretically any oil, no matter how high the titration, can be converted to biodiesel using the second stage base process after acid esterfication.

Biodiesel processed using the two-stage process tends to be a better product and provide a better yield of biodiesel. With low FFA levels, the extra yield is not really much, but the process itself is a lot cleaner and easier to convert. Less mixing time is needed, but more settling time will be required. Users in a rush to make biodiesel as fast as possible with low titration levels may not want to bother with the two stage process. Those with extra time and a desire to get the batch right every time with less soap problems should utilize the two-stage process.

First Stage (Acid Stage)

The first stage is performed by using sulfuric acid and methanol. The sulfuric acid should be mixed with the methanol prior to injection into the oil to be processed to avoid burning of the oil from the highly concentrated acid. Please be cautious when mixing: wear gloves, an apron/lab coat and goggles as the two chemicals can spit and splatter if mixed together too quickly. If too much sulfuric acid were to be mixed in at once, much like with a base catalyst, it could cause an exothermic reaction so great that the methanol could reach its flash point. Users should be very cautious when pouring the sulfuric acid into methanol and be sure not to let it over heat. It would take more sulfuric acid than is necessary in the process to cause a fire, but you can never be too careful with such dangerous combinations of chemicals. I personally like to use a bulkhead fitting in the top of the cone tank lid with a funnel so that the sulfuric acid can't jump straight back up through the funnel at my face.

The purity of the sulfuric acid should be at least 93%. The higher the purity, the better, since the impurity in sulfuric acid is mostly water. Concentrations as high as 98% can be obtained for reasonable prices. Any higher than 98% is unstable and will break down to 98% on its own naturally, so there is no point in seeking sulfuric acid more pure than 98%.

The amount of methanol to be used in the acid stage should be 12% by volume of the oil to be processed. It is difficult to say exactly how much sulfuric acid should be used, but there are some guidelines. One way is to perform a 1 liter test batch, using 1 ml of sulfuric acid per liter of oil, and to continue adding acid until the titration level no longer drops through the process. Titrations would need to be performed every few hours in the process in order to find the ideal amount of sulfuric acid to use for the main batch. For beginning users, this is a great way to formulate the general amount of acid needed for the type of oil being process. It's a lot of work, but the research can pay off for a single users case.

One rule of thumb is that titration levels will never drop below 2 times the volume of sulfuric acid added per liter of oil. For example, if 1 ml of sulfuric acid is used, the new titration of the oil will never drop below 2 x (1ml) = 2. Because of this ruling factor, it is good to avoid overshooting the amount of sulfuric acid required to fully treat the vegetable oil, otherwise more base catalyst will be needed to neutralize the sulfuric acid, and therefore more water and salt in the end product to remove from the biodiesel later. An undershoot of the amount of acid needed will result in more processing time and more base catalyst to convert. It is important to be as exact as possible to conserve resources, processing and washing time.

For those who wish not to spend time doing test batches for each main batch, a formula through trial and error has been developed by home brewers and shared on many forums. We've tested it and found that it works quite well. The formula is (Titration - Target Titration) x 0.2 = ml of sulfuric acid to use. The target titration is recommended to be set at a minimum of 3, since the creator of the formula had never gotten his titration to go below 3 in the testing and creation of the formula. This is because we are fighting acid with acid, and while one is removed, the other still needs to be neutralized. I have, however, been able to get lower than 3 on clean oil that actually starts off with an initial titration of about 3. By using 1 or 2 as a target titration with such clean oil, it's possible to get the lower target titration levels, close to the minimum titration governed by the x2 factor. Why bother with using the two stage process for such low titrations in the first place? The quality of the fuel is better, and there is a lot less mixing time required since the conversion is quick and flawless. I personally like the two-stage process because I can just mix in the acid and methanol and leave it to convert overnight. when i resume the next morning the base process only takes half an hour as opposed to 4 hours.

Start the acid process by mixing the acid and methanol together. Heat your oil up to about 55-60°C, then slowly dispense the mixture into the circulating oil just like you would in a base process. The mixture can be mixed in a little more ◻uickly than if you were doing the base process, but be sure not to mix too ◻uickly as you want to evenly distribute the methanol mix into the oil. Unlike in the base process, mixing during the acid stage is actually counter productive. Sulfuric acid tends to work faster without agitation. Once everything is mixed in well, shut off the mixing and let the sulfuric acid do its job on the oil. For those with stainless steel pressurized processors, adding more pressure and heat to really cook the batch helps speed up the process. The mixture is going to look like a dark brown mess once the methanoland sulfuric have been introduced into the oil. Have no worries as this is perfectly normal.

Titrations of the batch should be taken every hour to measure the drop in FFA content. Once the titration level becomes steady and no longer decreases after each test, the acid has finished and the 2nd stage of the process should be performed. One great thing about the two-stage process is the constant sample taking step is actually unnecessary. The process can actually be left alone overnight and the 2nd stage performed the next day.

Second Stage (Base Stage)

After the acid process is completed, the sulfuric acid and some water will either float to the top of the batch or sink to the bottom. If your titration was lower than 14, the water/acid will float to the top and not be economical to remove. Proceed to the next step for the 2nd stage of the process. If the titration level was above 18, the water and sulfuric acid will drop to the bottom, and some of it needs to be drained out. For the volume of fluid drained from the bottom, additional methanol will need to be added to the 2nd stage in order to gain a complete reaction. If the titration was between 14 and 18, it is up to the user to decide whether it's worth draining some of the acid/water from the bottom as some will float and some will sink.

Next, a new titration of the batch is required to perform the 2nd stage of the process. Mix the batch up a little before pulling a sample to be sure an average titration can be found. This titration determines how much base catalyst will be required to transesterfy the Biodiesel and neutralize the remaining sulfuric acid. The titration works the same as any WVO titration, however, since the volume of the batch has changed since the first titration done prior to adding the sulfuric acid, this new volume needs to be used for analyzing how much catalyst is needed. Use the titration number found for calculating the amount of catalyst needed to process the total volume of fluid in the processor, not just the original amount of oil used as in the first titration. This total new volume to base the amount of catalyst to use on will be the original volume of oil plus the amount of methanol added, so 1.12 times the amount of original oil processed minus any amount which was drained.

Add the amount of catalyst needed based on the new titration and new volume in the batch to the remaining methanol required to complete the process. Since the normal accepted value of methanol to be used is 20% of the volume of oil and 12% was used for the acid process, use 8% of the original oil volume for the remaining of the methanol plus any volume drained off from the bottom. Users who use more methanol than 20% can simply add to that 8% in this stage.

Mix the catalyst required with the remaining methanol. Be sure to add the catalyst more slowly than in the one stage process. Since there is less methanol to absorb all of the heat generated from this exothermic reaction, the methanol temperature will be higher than usual. Depending on the amount of catalyst needed, some cooling time may be required prior to dissolving all of the catalyst into the methanol. My preference is to add about half as much catalyst as will be needed right after the sulfuric acid has been introduced to the oil, thus giving the methanol plenty of time to cool down before I add in the rest. This allows enough time for the methanol to cool off before you top it off with the remaining needed catalyst. If your titration is still rather high and it is difficult to dissolve all of the catalyst into the methanol, add a little more methanol to allow a larger amount of catalyst to be dissolved.

Once the methanol/catalyst mixture has been made, begin to process the 2nd stage. Inject the methanol mixture into the oil slowly during the mixing of the oil. Once the methanol is fully mixed into the batch, give about 15-30 minutes of mixing and then perform a 3/27 test to check for completion. In the two stage process, the base stage will complete much more rapidly than in a single stage process. Most users will find a complete reaction within 1 hour, when it is common for it to take 3 or more hours in a single stage process.

Note: Some users may notice a white textured goo which is hard to get out at the bottom of the methanol mixing tank on the initial injection into the oil. This is caused by left over sulfuric acid coming into contact with the base catalyst, thus forming a salt. To prevent this, it is good practice to run a little wash of methanol into the batch once the tank has been fully drained, and then wipe it clean with a towel, or to just use two mixing tanks, one for sulfuric acid and one for potassium hydroxide.

As with the single stage process, once the batch passes the 3/27 test, shut down the mixing and let the batch settle the glycerin to the bottom. Drain the glycerin as usual and proceed to washing.

Examples of Catalysts

In chemistry, a catalyst is a substance that causes a chemical reaction to happen in a different way than it would happen without that catalyst - for example, a catalyst could cause a reaction to happen at a faster rate, or at a lower temperature, than would be possible without the catalyst.

When the term catalyst is used outside of chemistry, it refers to something that causes a change or reaction to take place.

Chemical Catalysts

Hydrogen peroxide will decompose into water and oxygen gas. Two molecules of hydrogen peroxide will produce two molecules of water and one molecule of oxygen. A catalyst of potassium permanganate can be used to speed up this process. Adding potassium permanganate to the hydrogen peroxide will cause a reaction that produces a lot of heat, and water vapor will shoot out.

The catalytic converter in a car contains platinum, which serves as a catalyst to change carbon monoxide, which is toxic, into carbon dioxide.

If you light a match in a room with hydrogen gas and oxygen gas, there will be an explosion and most of the hydrogen and oxygen will combine to create water molecules.

Intermediate compounds

In this process, a catalyst first combines with a chemical to make a new compound. This new compound is unstable, so it breaks down, releasing another new compound and leaving the catalyst in its original form. Many enzymes (special biological catalysts) work in this way. Many industrial chemical processes rely on such catalysts.

One example of a catalyst that involves an intermediate compound can be found high in the Earth's atmosphere. Up there, the chemical ozone (with molecules containing three oxygen atoms) helps protect the Earth from harmful UV radiation. But also up there is chlorine, which gets into the atmosphere from chemicals (chlorofluorocarbons, CFCs) used in some refrigerators, air conditioners and aerosol cans.

Chlorine is a catalyst, which steals an oxygen atom from ozone (O_3) leaving stable oxygen (O_2). At the same time, it forms an unstable intermediate chlorine-oxygen compound, which breaks down to release its oxygen. This leaves the chlorine free to repeat the process. One chlorine atom can destroy about a million ozone molecules every second. This can have a drastic effect on the atmosphere's ability to protect us from UV radiation.

Catalyst	Reaction
1. Manganese(IV) oxide, MnO_2	Production of oxygen from heating potassium trioxochlorate(V), $2KClO_3(s) \rightarrow 2KCl(s) + 3O_2(g)$
2. Vanadium(V) oxide or platinum (powder)	Oxidation of sulphur(IV) oxide to sulphur(VI) in the contact process for the manufacture of H_2SO_4, $2SO_2(g) + O_2(g) \rightleftharpoons 2SO_3(g)$
3. Reduced iron (powder)	Haber process - manufacture of ammonia, $N_2(g) + 3H_2(g) \rightleftharpoons 2NH_3(g)$
4. Manganese(IV) oxide, MnO_2 or platinum powder	Production of oxygen from decomposition of hydrogen peroxide, $2H_2O_2(aq) \rightarrow 2H_2O(l) + O_2(g)$
5. Nickel	Complete reduction of ethyne to ethane by hydrogen, $C_2H_2(g) + 2H_2(g) \rightarrow C_2H_6(g)$
6. Zymase (enzyme - organic catalyst)	Conversion of sugar into alcohol, $C_6H_{12}O_6(aq) \rightarrow 2C_2H_5OH(aq) + 2CO_2(aq)$
7. Diastase (enzyme)	Hydrolyses of starch to sugar, $2C_6H_{10}O_5(aq) + H_2O(l) \rightarrow C_{12}H_{22}O_{11}(aq)$

CONCLUSION

Through a process called transesteri cation, vegetable oil is refi ned with an alcohol in the presence of a catalyst to produce a fuel called biodiesel. Biodiesel is cleaner burning, nontoxic, and safer to use than petroleum-based diesel. Biodiesel meets the criteria set forth by the department of energy and is a registered alternative fuel with the EPA.

Biodiesel is known to be less toxic than table salt and biodegrades as fast as sugar. Biodiesel burns more completely, reducing CO_2 emissions by up to 75%. Biodiesel has a neutral net carbon output which helps to fight global warming. Biodiesel is proven to reduce the output of green house gases and reduces the need for foreign oil.

Biodiesel has demonstrated similar power ratings and fuel efficiency to petroleum-based diesel. Biodiesel is a powerful solvent (cleaner), which helps to keep tanks, fuel lines, injectors, and other engine components clean. Biodiesel can be mixed with any amount of petroleum-based diesel to create a biodiesel blend without any modifications to your vehicle.

Biodiesel can be made with the BioDiesel Kit for less than 70 cents per gallon. Other than the cost savings, biodiesel can also help our country's economy by reducing our dependence on foreign oil. Making biodiesel can also help our local farmers who produce oil bearing crops

www.ingramcontent.com/pod-product-compliance
Lightning Source LLC
Chambersburg PA
CBHW051311220526
45468CB00004B/1300